ceramics
a green approach

HERBERT PRESS
Bloomsbury Publishing Plc
50 Bedford Square, London, WC1B 3DP, UK
Bloomsbury Publishing Ireland Limited,
29 Earlsfort Terrace, Dublin 2, D02 AY28, Ireland

BLOOMSBURY, HERBERT PRESS and the Herbert Press logo
are trademarks of Bloomsbury Publishing Plc

First published in Great Britain 2025

Copyright © Kevin Millward, 2025

Kevin Millward has asserted their right under the Copyright, Designs and Patents Act, 1988,
to be identified as Author of this work

All rights reserved. No part of this publication may be: i) reproduced or transmitted in any form, electronic or mechanical, including photocopying, recording or by means of any information storage or retrieval system without prior permission in writing from the publishers; or ii) used or reproduced in any way for the training, development or operation of artificial intelligence (AI) technologies, including generative AI technologies. The rights holders expressly reserve this publication from the text and data mining exception as per Article 4(3) of the Digital Single Market Directive (EU) 2019/790

Bloomsbury Publishing Plc does not have any control over, or responsibility for, any third-party websites referred to or in this book. All internet addresses given in this book were correct at the time of going to press. The author and publisher regret any inconvenience caused if addresses have changed or sites have ceased to exist, but can accept no responsibility for any such changes

A catalogue record for this book is available from the British Library

Library of Congress Cataloguing-in-Publication data has been applied for

ISBN: PB: 978-1-7899-4194-4; eBook: 978-1-7899-4195-1

2 4 6 8 10 9 7 5 3 1

Edited and designed for Herbert Press by Plum5 Ltd

Printed and bound in China by C&C Offset Printing Co., Ltd.

To find out more about our authors and books visit www.bloomsbury.com and sign up for our newsletters
For product safety related questions contact productsafety@bloomsbury.com

THE NEW CERAMICS

ceramics
a green approach

Kevin Millward

HERBERT PRESS
LONDON · OXFORD · NEW YORK · NEW DELHI · SYDNEY

CERAMICS: A GREEN APPROACH

CONTENTS

INTRODUCTION 6
1. CLAY, BODIES AND FIRING 16
2. HOW RAW MATERIALS ARE
   PROCESSED INTO A CLAY BODY 28
3. RAW MATERIALS AND OXIDES 34
4. COLOURS AND OXIDES 40
5. FUEL FOR THOUGHT 50
6. KILNS AND POWER REQUIREMENTS 60
7. ATMOSPHERES 68
8. FIRING 74
9. THE RIGHT STUFF 80
10. FIFTY SHADES OF BROWN 90
11. BURNERS 102
12. GLAZE APPLICATION 106
13. PACKING AND DISPATCHING 118
14. HOW CLEAN IS OUR FUEL? 122
15. POTTERS 130
CONCLUSION 212
GLOSSARY OF TERMS 216
ADDITIONAL REFERENCES 220
WEIGHTS AND MEASURES 222
ACKNOWLEDGEMENTS 224

CERAMICS: A GREEN APPROACH

INTRODUCTION

Making pottery has never been more popular, as a hobby, a life style change or a possible career. Sadly the route into this profession has never been more difficult as a result of the decline of ceramics courses in schools, colleges and universities.

In general though, people moving into the world of ceramics are more aware of their impact on the environment, and the potential damage we could be doing to our planet. Unfortunately, many potters are detached from the materials and processes used in ceramics; from the preparation of clay bodies or preparing their own colours slips and glazes, to the placing of their pots into a kiln and the firing of their work. Recently I was lecturing on an MA course and I was surprised to find many students did not pack or fire their own kilns and were heavily reliant on brush-on glazes as opposed to glazes they made up by themselves. Although there is nothing wrong with off-the-shelf glazes, a certain degree of knowledge has dwindled to the point where even the use of the correct terminology is beginning to disappear. Worse, instruction in some settings is being given by people

Left and title page: Me on the potter's wheel through the ages! Photos © Kevin Millward

with little or no training in the profession. As an example, someone attended one of my beginners' throwing courses and was most insistent that she should be able to throw pots by the end of the two-day course. I explained she would be able to centre the clay and form a basic cylinder by then but she would not have mastered throwing. When I asked why she was so concerned she said: "You don't understand! I have spent a lot of money setting up a pottery studio, and I have a full class on Monday, teaching them how to throw pots". Amazing.

Mid-eighteenth century pots that inspired me as a child at our local museum. Photo © Kevin Millward

INTRODUCTION

As potters we deal with a wide variety of materials and processes that all impact the planet in some way. Given this it is important that we understand where they all come from and the journey they have made to our studios so we can manage them responsibly. My journey as a potter began over 50 years ago and my experience has given me a unique perspective to give an overview of the green credentials of both studio and industrial ceramics and how the lessons learned over hundreds of years are still relevant today. I constantly see potters unaware they are 'inventing' solutions to problems that were solved hundreds of years ago.

My introduction to pottery started in the late sixties at my high school's pottery department. I built a slab pot and sold it to my brother who took it with him to university. On leaving high school I enrolled at art school with the intention of being a sculptor but on entering the pottery department I was drawn to the potter's wheel and, working with clay again, my lifelong obsession with ceramics began. The move away from sculpture was a big decision as it was in the blood; my great grandfather was a sculptor whose chosen material was stone. In moving to ceramics it was immediately evident to me that the first hurdle was acquiring the skills necessary to make the pots I envisioned, let alone the technical aspects of all the materials needed to colour and glaze the pots, then fire them. Everything required to make your pot was available with just a trip to the store room. The clay, glaze materials and use of the kilns, both electric and gas, were all free, as was the four-year course. It was only after my introduction to clay that I realised I'd always been fascinated by ceramics. As a small boy I would accompany my mother on archaeological digs organised by Dr Francis Saloria, Head of Archaeology at Keele University, and watch the expression of amazement as they found small scraps of pots in holes in the ground. These small pieces of pottery were often then kept in brown cardboard boxes under my bed. I loved trips to the local library and its small museum. Although it mostly celebrated the town's history of silk production, Leek had a close association with William Morris. One of the families he stayed with had collected mid-eighteenth century ceramics and these were donated to the town and are now on display. Cases full of teapots in Rockingham and

Whieldon ware glazes fascinated me long before my introduction to clay.

It has always been the whole spectrum of ceramics that has interested me including the design of table ware, designing and building ceramic equipment and kilns.

After leaving art school in 1973 I went to work in a small studio pottery in Cheshire called Coopers Pottery producing domestic stoneware items. The studio had the biggest kiln I had ever seen – you could have put all the art schools kilns inside this one; they also had a small salt kiln. It would take all morning just to brick up the door. It was fired with forced air oil burners for efficiency. This was, in some ways, was my first introduction to maximising the use of fuel in firing, whether it be electricity or oil. The learning curve was steep, working with both art school trained potters and ones who had completed the pottery managers training course at Staffordshire polytechnic – the same one attended by David Leach in the 1930s before he returned to St Ives to help set up the pottery there. I also had the privilege of working with Leach for short periods of time when he visited old friends in Stoke.

My work at Coopers involved producing a full range of domestic stoneware biscuit, fired in electric kilns and glaze fired in the oil kiln. I then moved on to the newly opened Gladstone Pottery Museum that was setting up a small studio where they produced terracotta domestic and garden ware. This is where David Rooke, the lead potter, taught me in the art of throwing large pots. He'd learned this skill at Brannams in Barnstable, an old established country pottery. The skills of production throwing, involving the throwing of pots ranging from 227 g (8 oz) to 27 kg (60 lb) was a game changer for me. It's a skill that has stood me well over the years and I can still throw a small flower pot in under 50 seconds. I was introduced to the process of industrial ceramics by John Gould, who had worked as a pottery manager for some of the largest producers of earthenware and bone china in Stoke-on-Trent. His expertise in setting up glazes and slips, the use of pint weights, torsion viscometer flocculants and deflocculants and problem solving was inspiring. All this alongside other potters who had worked in the industry; a place steeped in the history and techniques of making, glazing, and onglaze decoration and firing used in the Staffordshire pottery industry, both past and present. I even got to see a bottle oven fired on coal. The world of ceramics has changed so much in such a short space of time – we have gone from firing kilns manually and forgetting to turn them off (as I know to my own cost; this is how I learned to wind my own elements), to basic clockwork methods of controlling kilns and turning them off and now to fully digital control systems you can set from your smartphone. Another mentor was Derek Royal, who worked with the great Emmanuel Cooper on most of his glaze books.

It was my knowledge of both studio and industrial practises of control of materials and processes to make ceramics that led to the approach by a Stoke-on-Trent-based company asking me to go to Ireland to give a demonstration on throwing big ware. After the success of this seminar I was asked if I would work full time for Harrison Mayer as a craft and technical advisor. Apparently they had been trying to fill this post for a while but could not find anyone who had both the practical skills in making, and the technical knowledge of both studio and industrial ceramics. Harrison Mayer were world leaders in the supply of ceramic materials and equipment to both the industrial and craft sectors, as well as in the production of colours, glazes and preparation of raw materials for companies such as Wedgwood and Royal Doulton, to name

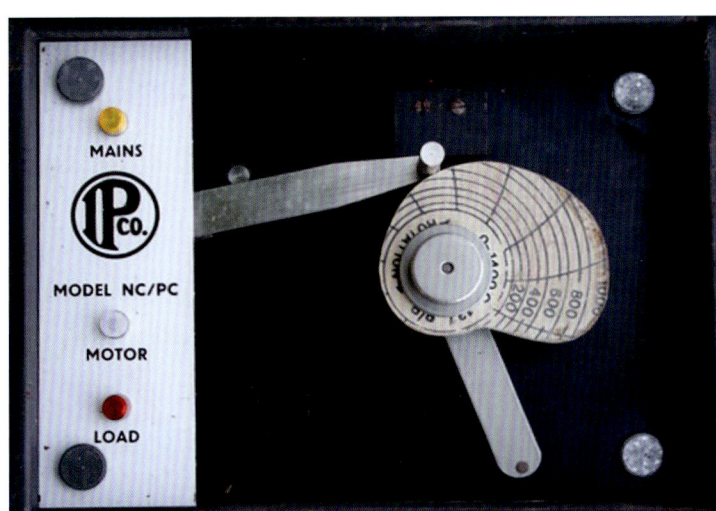

Mechanical cam controller as used by me in the seventies on electric and gas kilns. Photo © Kevin Millward

but a few. They wanted to break into the craft and education market and expand the hobby sector, and my job was to help facilitate that. Better to have a friendly face with a background in studio pottery than a lab technician in a white coat. We also wanted to compete with small companies whose material consumption was too minor for the parent company to deal with. This sector was already dominated by some well know companies such as Podmores, Wengers and Potclays who were well connected with studio potters. Harrison Mayer had the finances to take on these companies and hopefully dominate the market. It was during this period I was tasked to explore the world of American hobby ceramics, testing American-produced potters' wheels, top-loading kilns, brush-on glazes, casting slips and preprepared colours in small containers to bring to market. With all these resources the market exploded. For the first time potters could buy a small kiln that could be fired at home with a potter's wheel small enough to tuck away in a garage or shed; brush-on glazes and colours ready to use at low cost and attainable with little or no technical knowledge required. This led to a whole host of problems as potters tried to use the US and UK systems together leading to many problems, some of which still haunt us to this day. It became my job to help sort it out.

While working there I came into contact with some of the industry's greatest minds. They had worked out how to solve most of the technical and making problems that had plagued potters, using their knowledge of ceramic materials and processes. I find it sad, if not amusing, to see potters wrestling with a technical problem unaware that the problem was solved and the solution found over 200 years ago in the industrial sector. As the two sectors have never really communicated with each other, I have tried to bridge this gap over the years, given that more and more potters are adopting industrial materials and methods of making into their practice with little or no information on how to use them correctly.

It was while working with Harrison Mayer that it became very clear to me that the industry's response to the use of materials and its insatiable demand for fossil fuels had led to the development of more efficient ways to save fuel. Meanwhile the craft sector has, in many cases, been left behind; burning a wide range of fuels in less than efficient ways – mostly because fuel used by studio potters was, in the past, relatively inexpensive.

After leaving Harrison Mayer I set up a studio in my home town of Leek. Taking advantage of the knowledge acquired from both the craft sector and industry I built a 0.99 m^3 (35 ft^3) reduction gas kiln with forced air burners, maximising the efficient use of natural gas. It was at this time my friend Greg Daly was over from Australia. We fired many of his pots in my kiln, and he was amazed at its speed and efficiency as it fired in about eight hours to cone 11 with perfect results. To this day this is a rare thing to see on studio kilns, but standard for industrial kilns. I also started teaching at my old art school and in many of the leading universities that specialised in ceramics in the UK. The creativity was off the scale but the lack of technical knowledge in some of the ceramic departments was astonishing. Ceramic processes such as slip casting, firing and glazing to a finished piece were often missing. There was a lot of waste, especially when it came to the glazing and firing, due to the lack training and expertise in these areas; the failure rate was invariably massive – lots of pots stuck to kiln shelves – most of which was avoidable. Teaching students these technical techniques is essential.

The next stage of the material journey is its use and processing into a finished object and the disposal of the waste. The industry has, for many years, had to conform to strict guidelines for the use and disposal of all ceramic waste materials from clay, glaze, colours, plaster moulds and broken ceramics, but some people involved in pottery give little or no thought as to what they flush down the sink into the water course. At Harrison Mayer all water use in production was cleaned by filter pressing, thus enabling the proper disposal of solid waste ceramic materials and only discharging clean water. How many potters have you seen flush ceramic waste down a sink or toilet? I think that the time is right to rethink how people who practice ceramics can respond to a greener approach, one that allows the potter to be as creative as possible without having a detrimental impact on the environment. Hopefully this could be the start of a green revolution for ceramics in the craft sector.

Finally, as well as my work in the craft sector I have been heavily involved in TV, bringing ceramics into many programs, game shows and films – most notably *The Great Pottery Throw Down* which I developed with Love Productions. I have also designed and developed some of the best-selling industrial manufactured table wares in the UK and USA and one of the best-selling clay bodies, KGM, with Valentine Clays as well as designing and building kilns and ceramic machinery. With over 40 years' experience of teaching on BA and MA courses, this then culminated in the setting up of Clay College with Lisa Hammond and Adopt-a-Potter that is based in Stoke-on-Trent, the historical home of ceramics in the UK. This was set up with the intention of training the next generation of potters and having one eye on the environmental impact as well as the most efficient use of materials and fuel.

A friend of mine once said if you want to be green don't be a potter! As we progress into the twenty-first century the impact that creatives have on the environment we all share becomes more important. As potters we are very much involved in the last stages of the raw materials' journey into a finished object. It is very easy to only think about what it will look like when your work emerges from the kiln and not what the total impact has been on its complex route to the studio and the stages of its creation. Many potters are concerned, but while

Used plaster moulds going for recycling. As controlled waste it is kept separate and is used in the production of plaster board. Photo © Kevin Millward

crushing up a few broken biscuit pots and making your own grog is laudable it is not the answer in the long term. Potters should not be totally ignorant of the global impact of the craft – we need to understand our place, no matter how small, in a worldwide movement towards cutting unnecessary waste and making art more sustainable. Unfortunately there are still hobby and professional potters who give little or no regard to where their materials come from, how far they have travelled, what level of processing is involved and the environmental impact of the extraction and preparation involved in bringing these to the studio.

I hope this book will be a journey of exploration into how potters can be more aware and responsible in the way we source and use the planet's finite resources. I aim to help find the most effective way to use the chosen fuels to fire our work and those which will give the best results both aesthetically and practically. We must also not forget that the finished work has to be packed and dispatched, or even personally delivered, which has an environmental impact on road, sea and air miles. How can we achieve this with the least possible ecological damage, leaving a legacy for future generation as a mark of our creativity, and a reflection of our respect for our planet and civilisation?

Given the nature of all the different materials, fuels and processes a potter has to deal with during their practice, how can we be as green as possible? The start of this journey is to understand what we are working with. *Ceramics: A Green Approach* is no way intended to be a detailed instruction manual or academic research paper quoting facts and figures, nor is it instructions on how to achieve a particular effect or savings; it is to start a conversation. I aim to introduce the reader to what could be the best way forward for our craft, and to help potters to be more aware and informed. What are the effects the ceramics craft has on the environment? And how do we lessen its possible detrimental impact upon it? I hope this book will help with these questions.

CHAPTER 1

CLAY, BODIES AND FIRING

As surprising as it sounds, pots are not made from clay; they are made from 'bodies'. These bodies are closely linked to the temperatures they are fired at and determine their classification.

Most clay dug straight from the ground is not viable without some form of processing to make it usable in pottery. Many potters dig and prepare their own found clays into viable bodies. The blend of different clays, plus other raw materials, constitutes a 'body'. You can read more about this process in a book called *Wild Clay* by Matt Levy, Takuro Shibata and Hitomi Shibata.

Some bodies only contain 25% of a natural occurring clay. For example, bone china consists of 50% calcined cow bone, 25% china clay, and 25% Cornish stone. Bone china was developed by Josiah Spode in the mid-eighteenth century in Longton, Stoke-on-Trent, as a competitor to European porcelain.

China clay extraction. Cornwall has been the main producer since the early nineteenth century in the UK. Photo © Getty images

Eighteenth century European Porcelain, an example of pure white vase often referred to as white gold due to its cost of production and retail value. Photo © Getty images

This was developed alongside white earthenware and creamware bodies and was the driver for the development of the china clay industries in Cornwall, specifically in St Austell, enabling the development of white firing bodies.

William Cookworthy is credited with developing one of the first hard paste porcelain bodies in the UK in around 1755.

Throughout the history of pottery, potters found it easiest to use the most commonly occurring local clays that could be easily extracted from the ground with minimal preparation. Colour and glaze were to come much later.

One of the main driving forces in the refinement of ceramics in Europe was the importation of porcelain from China. The word 'porcelain' was invented by the Italians, as the colour and translucent quality of the imported pottery reminded them of the cameos cut from the porcellanous shells.

The high quality of these ceramics created a market for fine ceramics across Europe, at such high prices that it was referred to as 'white gold'. It created a ceramic revolution in bodies and the raw materials, and in the methods used, in what would become the industry of ceramics.

In the UK, it initiated the extraction of china clays, ball clays, fire clays and red clays, for a wide range of applications within both domestic and industrial potteries. Bricks and tiles were able to be manufactured from these materials.

Many studio and hobby potters buy their clay from a supplier with little or no understanding of the lengths it has taken to get it into the plastic bag that they take home with them. If there was no industrial use for these raw materials, they would not be available for common purchase.

The close association between clay bodies, the potter, and the potter's fuel source has long been problematic, as the refinement of the product often meant that the raw materials, the skilled labour and the fuel were not all in the same location. In the past, potteries would be set up where the natural resources occurred, Stoke-on-Trent being a perfect example of this. Clays, flint and feldspars from the south, lead from nearby Derbyshire, and coal from

North Staffordshire, all came together to create the area known as 'The Potteries'.

The pottery industry today operates on a global scale. The importation of raw china clays in plastic form, from as far away as New Zealand, has increased due to the demand for a porcelain body that mimics the looks of bone china. The movement of raw materials has always been an issue, due to the weight. Pack mule was the beast of burden at the beginning, bringing the vital components to Staffordshire. Further development of the canal systems developed by Josiah Wedgwood and James Brindley allowed materials to be transported into Stoke and the export of the finished goods out into the world.

The environmental impact in today's modern world is considerable, given the amount of energy and water involved in the extraction of clays and raw materials. This contributes to rising prices for potters, also increased by transportation costs, which have never been higher.

A potter once asked me how the price of clay could be justified, when it is just dug out of the ground and put in a plastic bag. Many potters have little or no understanding of how clay bodies are produced and often will not understand the categories for the different bodies, and the temperatures required for firing different bodies, which can lead to unnecessary waste.

The following sections look at the different types of clay bodies commonly produced for the UK market.

EARTHENWARE

Earthenware clays are porous and are normally low fired at under 1200°C (2192°F). Most commercially available clay bodies are normally either white or cream in colour, based on china clays and white firing ball clays, or they are sedimentary red clays, i.e., terracotta. It is normally acknowledged that the whiter the clay body, the less plastic it is. When some plasticity is required some of the whiteness is sacrificed. Industrial manufacturing, where moulds are used, often requires a more plastic body.

One aspect of white earthenware bodies that has led to confusion is that very often the maturing temperature is in fact higher than many earthenware glazes.

Industry uses higher biscuit temperatures that the glazes can withstand. Commonly the biscuit temperature would be about 1140°C (2084°F) with a glaze firing of about 1080°C (1976°F). This can cause confusion over how a glaze can be added to the pot without it being in a porous state. However, you don't need porosity to apply a glaze. You don't need to warm the pot, as potters have developed ways to allow the application of glaze to vitreous ware for hundreds of years. I will cover this more in the chapter on glaze application.

Many potters working with these earthenware bodies could not, or would not, adapt to these methods. Instead, they would opt to use lead-free glazes that can be fired to 1140°C (2084°F), enabling body and glaze to mature together. If the appropriate firing processes are not observed then the glaze will be affected, making the pot non-viable as a functional item.

Over firing the glaze can cause blistering and other problems, while under firing can lead to crazing; both are a common cause of loss in earthenware pottery. Knowledge of the materials and correct processes can save a lot of waste in this type of ceramics. Most potters using terracotta will biscuit fire at 1000°C (1832°F) and glaze at 1080°C (1976°F) as the common maturing temperature is usually about the same. It is worth noting that many lead-free glazes don't respond well on terracotta and can turn an unpleasant colour, again leading to waste.

STONEWARE

Stoneware can be any colour and is a vitreous body, so will hold water once fired, even if not glazed. It is usually fired in excess of 1200°C (2192°F) with a feldspathic glaze. Most bodies are based on one or more clays – for example ball clays, fire clays, china clays and red clays. These are normally very good to throw with due to their plasticity. Choosing between a base of ball clay or fire clay will often come down to cost, but most potters would consider a ball clay base gives the best working qualities. In most cases these contain less iron so you can have a body that is less or more toasty in colour, based on your choice.

An example of coloured stoneware would be a Wedgwood blue vase, but these are the exception. When it comes to something that throws well, it may surprise you to learn that the best black clay bodies come from a white clay base. There is a stoneware body that is as white as porcelain but does not have its translucency. The range of stoneware bodies available today is vast and covers almost every possible variation in colour and finish to suit your firing style or glaze. Even with access to such a vast range of clay bodies, some potters still prefer to make up their own.

The cost of transporting tonnes of clay bodies in plastic form, with the water content adding to the already considerable weight, can be an incentive to transport lighter, dry, raw materials and then blend with water on site. Most raw clays are processed wet and are then dried out for bagging, using vast amounts of energy. The process used is similar to that used to make dried milk, but many companies use spray drying. This involves spraying the wet material into a heated rotating drum, rather like a giant tumble dryer. In the past they sometimes used metal trays in a big drying cupboard. If you were a real stickler, you could identify a tray-dried clay or glaze by the fact it was lumpy rather than a powder.

PORCELAIN

This is an area that is far more complicated than it appears. The perception of what porcelain should look like has changed over the years, mainly due to taste, the rise of the hobby potter and the top-loading electric kiln.

The search for porcelain has a long and fascinating history all of its own but to make it simple and pertinent to the studio potter, porcelain could be said to fall into two types. There are high-fired reduction porcelains, used by potters with kilns capable of reduction firing, achieving high temperatures of up to 1400°C (2552°F), with glazes associated with this style of pottery and the kilns – for example, celadon, chun, shino and tenmoku. This type is, on the whole, based on china clays from the UK and white firing ball clays which tend to produce a cream-coloured porcelain in oxidisation, owing to the small amount of iron present in the base clays. In reduction firings this small amount of iron tints the body into the blue-white spectrum, giving the classic qualities of oriental porcelain and not that of bone china.

The more common form of porcelain today is the super-white body that is aimed at the potter with an electric kiln, firing to lower temperatures of about 1240-1260°C (2264°F-2300°F) in oxidisation. This demands the development of ever more white bodies that are plastic enough to throw well, very often with the addition of bentonite to plasticise the body. Sometimes there is so much bentonite in the body, potters find they can't make a casting slip with them, in which case ask your suppler first to see if it is suitable for slip casting.

These, of course, have excellent translucency, even at lower temperatures. The definition of porcelain is white, vitreous and translucent, with a feldspathic glaze maturing together at a high temperature. I believe that the search for ever more white porcelains, that fire in oxidisation to a super-white, is driven by a desire to create something that looks more like bone china, which does not lend itself to studio pottery due to its lack of plasticity.

In terms of environmental impact, clay producers have to search far and wide

Reduction red porcelain bowl fired in small commercial manufactured propane gas kiln. Photo © Kevin Millward

for the whitest firing china clays to satisfy the demand for ever more white porcelains, some coming in plastic form from as far away as New Zealand. How can we justify this? It is interesting to note that many potters, in their quest for these porcelain bodies, find the reprocessing of scrap and waste very difficult, as the chance of contamination is very high. It can be difficult to deal with without the scraps needing to be slip housed, to remove any contamination, which is not viable for most potters; instead, it is often discarded.

The traditional forms of porcelain are still prone to iron specking, but some potters are happy to have this rather than sacrifice the loss of plasticity that occurs in the process of slip housing. Students of Clay College in Stoke have made up porcelain bodies using our 'Peter Pugger' pug mill with great results, with the increase in throwing qualities being notable improved.

BONE CHINA

This is probably the easiest body to describe, as to be called bone china it has to adhere to a fixed formulation. Using 50% calcined animal bone, 25% china clay and 25% Cornish stone, this recipe was defined by Josiah Spode in the middle of the eighteenth century. It was the English response to porcelain, as they did not want to give up on their tradition of high-quality, lead-based glazes. This led to the industry standard, white vitreous and translucent high fired biscuit at 1260°C (2300°F) and the application of a low glaze firing at about 1080°C (1976°F).

I will discuss methods for applying glaze to vitrified surfaces in the chapter on glaze application.

Bone china is one of the most expensive bodies to produce. Its production requires pure white cattle bone, usually imported as it must be from grass-fed cattle. This is calcined to remove any organics, then ground to a specific particle size, before being mixed with the other ingredients.

The level of cleanliness needed to achieve the desired whiteness is incredible, and it is not unknown for some producers to send their scrap back for reprocessing by their supplier to avoid cross contamination. Some potters may be offended by this particular body as it is not vegan or vegetarian due to its bone content. There could be an argument that this is a recycled body, as the bone is the by-product of the beef-farming industry, but that is still problematic for some people. The level of whiteness achieved due to the bone is unsurpassed; while the body is awful to throw with as it has little or no plasticity, it is ideal for slip casting or machine making using plaster moulds.

TIP: The good thing about all these body types is that they are all recyclable at any stage up until firing, irrespective of how long they are left standing. It is obviously much easier to recycle the clay before it dries out by wedging and kneading, but if it does dry out it can be slaked down with water, and with a small amount of effort and no specialised equipment returned to usable clay body, as long as the clay is completely bone dry. There is no need to break it up as it will slake down on its own overnight. Leather-hard clay does not always break down and can end up in lumps. Extra care should be taken with porcelain and bone china to avoid contamination with iron oxide, i.e., rust. If you are at all concerned about the possibility of contamination you can make this into a very thin slip and pass it through a sieve. You can put a non-ferrous magnet from an old speaker in the sieve to catch any loose iron particles. This would depend on the body type; grog or no grog 80s would be a good start, to remove any large form of contamination, especially any plaster or limestone. It is left to settle before the water is removed, leaving the clay in a slop state. It can then be put on dry plaster bats to draw out more water before it is ready to wedge or pug.

If your scrap has been reprocessed many times it will need to be rested for some time – up to, and over, a year. This is commonly referred to as weathering. It will make the clay better than new, and if the frost gets to it all the better. If this does happen, you will need to re-pug the clay before use. If you have concerns about your plaster bats, cover them with old nylon sheets. Nylon will allow water through and won't rot as fast as cotton; it will prevent bits of plaster getting into your clay. You can even put your clay slop in nylon pillowcases, suspend them and allow gravity to draw out the water. If you are going to re-use old plaster moulds then make sure the broken-up plaster is wet, as it will allow better adhesion with the fresh mix.

BASIC TYPES OF GLAZES ASSOCIATED WITH THESE BODIES

EARTHENWARE

Earthenware firing uses low temperature glazes, mainly in oxidisation, from as low as 1000°C (1832°F) up to just under 1200°C (2192°F). The glazes are normally based on frits as the main glass-forming element. Frits are made by melting the raw ingredients together in a kiln (for example silica and lead) that when molten, are poured into cold water, which is spectacular to watch. This creates a glass ash, locking in the toxic materials, such as lead, so that handling the glaze does not pose any risk. It is then ground down in a ball mill with water, using porcelain pebbles, until it is fine enough to use in a glaze. It is dried out and bagged to be used in conjunction with other raw materials, such as china

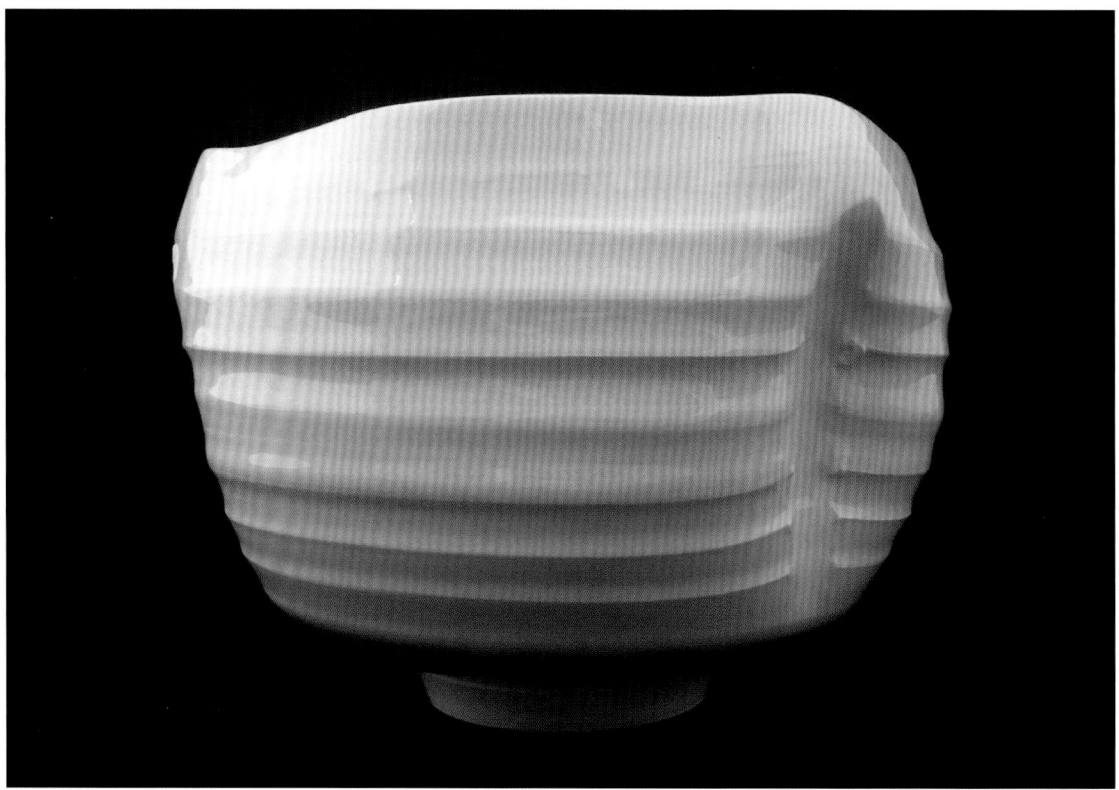

Cream earthenware bowl. Photo © Kevin Millward

clay, silica and feldspar, for making glazes. Borax and soda-ash-based frits are also processed in this way to render them insoluble. Large amounts of energy are used in the frit-making process, irrespective of what type of earthenware you are producing. This type of glaze is also used on bone china.

STONEWARE AND PORCELAIN

The range of bodies available is so wide it is impossible to list them all here, but it would be fair to say that irrespective of how the piece is made, most will be biscuit fired at about 1000°C (1832°F). The feldspathic-based glaze is applied, then glaze fired to about 1280°C (2336°F) in a kiln fired by oil, gas or wood; it will be slightly lower for oxidised firing in an electric kiln. Many surface qualities and finishes demand a whole range of different methods of glaze application. An assortment of surface effects could include ash landing on the pot in a wood-fired kiln, glazes made from rocks, or ash produced using a wide range of plant materials.

Salt or soda firing is very popular providing you have a suitable location in which to introduce the salt or soda into the kiln during its firing. The main environmental problem with this practice is that toxic fumes from the kiln produce acid rain, in addition to localised chlorine gas around the kiln. These processes are banned on an industrial scale under the Clean Air Act, but due to the very small number of potters involved in such firings, the amount of pollution is minimal and, in some cases, can be justified by the high quality of the work produced by excellent potters.

Whether or not to use this type of firing is therefore down to the potter's personal conscience, unless a higher authority decides to ban it.

TIP: Remember that your glazes, other than brush on, should be sieved every time before you use them, to prevent transferring any debris in the glaze onto your pot. 100s mesh is normal for this, after which you can then set the pint weight.

Brushing your biscuit-fired pots before glazing is good practice. I don't approve of the sanding and washing of biscuit pots for many technical and health reasons. It is better practise to get them right at the clay stage, before you fire them.

CHAPTER 2
HOW RAW MATERIALS ARE PROCESSED INTO A CLAY BODY

SLIP HOUSING

The main method of processing clay in the UK is called slip housing and was developed as a response to industry's demand for less iron contamination in its bodies. It is sometimes referred to as the 'English method'.

In this process all the materials are mixed individually with water and passed over powerful magnets to remove any iron in the form of rust, then passed through fine sieves to remove any oversized particles. It is not common knowledge that the ingredients for the body are blended wet, a crucial factor to allow controlled dust-free blending of the ingredients. The resulting slip is not deflocculated, so the blunger cannot be stopped as the slip would settle out. Large amounts of clean water are used in this process.

Pan milling. Potclays Ltd is one of the only companies to use this method. Photo © Ben Boswell

CERAMICS: A GREEN APPROACH

Filter pressing clay body, Valentine Clays

Some country potters in the past would use the waste heat from their kiln to sweat out the water, but today we use a filter press which basically squeezes the water out through fine cloths, preventing the clay particles from passing through, but allowing the water to do so. As you can see, a large amount of energy is used just to get it to this stage.

Now the clay body can be removed from the filter press. At this stage it is referred to as filter cake. It should be noted that at this stage you can use filter cake to make casting slip with no need to pug it, thus saving unnecessary time and energy pugging and bagging the clay. The next stage is to pug and de-air it through a vacuum chamber before extruding it into the correct shape, weight and hardness, after which it can be bagged. You can specify the hardness of a clay body depending on what you want to do with it. This is critical for industry as certain processes demand the body to be in the correct state for use on the appropriate machine. This is the same for studio potters, as they require clay to be in a suitable state of softness or hardness, depending on the specific project.

This method of manufacturing clay bodies is a costly process, and it may remove some of the plastic qualities so revered by potters, unless they are willing to set the body aside for a length of time (referred to as weathering or souring) to improve its plasticity.

PAN MILLING

Another method is dough mixing or pan milling which is less common today, as it is a more hands-on method where potters who make their own bodies use an old dough mixer.

Pug milling clay body, Valentine Clays

Potclays, a supplier in Stoke-on-Trent, use pan mills formerly used in chocolate factories. The raw materials are usually fed dry into the mixer from 25 kg (55 lb) bags. This clay has already been dried out once before. Water is then fed in until the desired hardness or softness is achieved and can then be removed, after which the body can be pugged and de-aired before being packed into plastic bags.

From an environmental angle, sadly it is not practical for producers to re-use the plastic bags, due to contamination, damage to the bags, and the amount of clean water it would take to wash them. It is better for the potter to re-use them or dispose of them responsibly themselves. Clay packed in Europe comes in vacuum-packed plastic. Prior to plastic, damp hessian sacking was used to cover the clay, as the clay was produced and stored on site, but it is not suitable nowadays in terms of long-term storage in warehouses and studios, as well as transportation over distance.

The advantage of the pan milling method is that it maintains the plastic qualities of the body and uses far less water and energy. The downside of pan milling for some potters is that small amounts of contamination from iron or other impurities can find their way into the body, most noticeable in porcelain bodies. A whiter, cleaner porcelain body is always created using the slip housing method, but it won't throw as well as one that is pan milled.

How does this affect the green-minded potter? Unless you are willing to process the clay body yourself – as many potters have done in the past and some still do today – you will need to buy your bodies from one of the main suppliers, and you will have to accept all the financial and environmental costs of the water and energy used to process and bag it in plastic and ship it to you by road. The cost of transport nowadays is becoming an increasing concern, so perhaps potters should be thinking about increasing the quantity they have delivered, if there is room to store it, as this is more cost effective in both fuel costs and getting a better price on the clay.

Now to the problem of reprocessing scrap. As discussed, scrap can be reprocessed ad infinitum, as until it is fired it can be reclaimed. Once you have set up a system for reprocessing, it is not difficult, even if you don't have a pug mill. You need to keep on top of it though, otherwise it can become a big problem. In my first teaching job at my old art school, I made an effort with some of the students to reduce the amount of scrap, as we did not have a full-time technician. I was surprised to find a pot in the bottom of a scrap bin that I had made when I was a student some ten years earlier!

It may be considered cheaper just to throw the scraps away, as the time taken to reprocess them may seem more valuable than the cost of the clay, but this is wasting a valuable, recyclable resource. Some manufacturers are afraid of reintroducing scrap back into the system because of the outside chance of contamination, so they feel it is more cost effective to discard it. The processing of scrap can also drive you to improve your making because if you are a competent maker, you should not have too much scrap anyway. Too much turning to improve badly thrown pots is not good craftsmanship; remember turning is for refinement and enhancing a form, not for the removal of excessive weight.

There may come a point in the not-too-distant future where it becomes viable to produce your own body again. This would obviously still depend on having space for machines and storage for large quantities of raw materials, as many potters today are more urban and have small studios. If you are super keen you could look at digging some of the raw clays yourself. I always do a dig-your-own clay project with my students, and many incorporate it into standard bodies or glazes, so customising a clay body with a unique identity.

TIP: The protocol for testing your own dug clay involves drying out the clay, slaking it down with lots of water, sieving out the debris, returning it to plastic clay, then making small tile marks of 10 cm (3.9 in) on them, before firing at a range of temperatures, up to stoneware and reduction. Monitor the results and you can work out the shrinkage rate and any fluxing of the clay. Some could make great glazes; others may be very good for making a clay body or adding to a body. It's a good idea when high firing to put them in a small saggar, as they may melt.

CHAPTER 3

RAW MATERIALS AND OXIDES

If you want to be as green and carbon neutral as possible, it is important to know whereabouts in the world your raw ingredients come from and under what conditions and circumstances they are extracted from the earth. Look into the effects this has on the local population and environment and what other journeys are taken along the production line, before they are ready for use. It is interesting to know that the main consumers of these materials are industries such as paper making and construction, and it is very rare that they are produced specifically for use in ceramics.

Many potters are rightly concerned about the impact ceramics has on the environment and many have made moral and environmental decisions on what they choose to use and how they fire their pots. We make many such choices in our everyday lives – whether to buy free-range, organic eggs and traceable home-grown fruit and vegetables, not using plastic bags. Do we apply the same consideration when we buy our raw materials and clays? It was only a few hundred

Granitic Mylonite pink bands containing feldspar. Photo © Getty images

years ago that the colour and form of a potter's work would have depended on which clays, raw materials, fuels and oxides were locally available. In a modest way this chapter covers where our materials come from and who prepares and reprocesses them; you may be surprised.

Flakes of rust iron oxide used for additions to glazes and bodies. This came from iron work on an old gas kiln. Photo © Kevin Millward

MAIN PRODUCERS ON A GLOBAL SCALE?

TIN OXIDE: China

LEAD: China, Australia and United States of America

COBALT: Democratic Republic of Congo

MANGANESE DIOXIDE: South Africa, Australia, China and Mexico

COPPER OXIDE: United States of America and Chile

IRON OXIDE: Australia, Brazil and China

CHROME: South Africa and Zimbabwe

TITANIUM: Australia, South Africa, Russia, Sierra Leone and Japan

VANADIUM: China, South Africa and Russia

CADMIUM: China, Japan and Korea

SELENIUM: Japan, Canada and United States of America

IRON CHROMATE: South Africa and India

ILMENITE: South Africa, India, United States of America, Canada, Australia and Russia

DOLOMITE: United States of America

POTASH FELDSPAR: United States of America

SODA FELDSPAR: United States of America

NEPHELINE SYENITE: Canada

WHITING/LIMESTONE: England, France and Belgium

SILICA QUARTZ FLINT: Japan, China and Russia

TALC: United States of America

CORNISH STONE: England. No longer commercially available but substitutes are available

CHINA CLAY: England

BALL CLAYS: England

FIRE CLAYS: England

FFF FELDSPAR: Finland

COLEMANITE: United States of America and Turkey

LITHIUM: Chile, Australia, Argentina, and China

BARIUM: United States of America, China and India

These materials may be found in many other counties in lesser quantities.

The list below shows the approximate percentage of global market share for the preparation of glazes and colours.

SPAIN	40%
CHINA	10%
U.S.A.	10.5%
INDIA	10%
ITALY	8.5%
NETHERLANDS	6.5%
GERMANY	5%

Although the materials are obtained from these countries, it is very unlikely they are reprocessed there for use in ceramics because, as you can appreciate, the materials themselves are used in many other industries.

At one time the UK was a major manufacturer of colours, glazes and clay bodies for both its domestic ceramics industry and for export, but it is now no longer a major player. It may surprise many to find that Spain is one of the top processors of ceramic materials.

Very few potters are aware of the distances involved in the transportation of raw materials and the large quantities of energy and water used to produce the plastic barrels, underglaze and onglaze colours, stains, oxides and glazes used every day by both hobbyists and professionals alike. Many of these materials will have travelled the world before they find themselves on the shelf of your local ceramic supplier, very often in plastic pots and bags. Glazes are made wet and supplied wet for industry but are dried out for small producers, which often doubles the price of the glaze.

One of many concerns is what the alternatives would be should large industry no longer have a need for these materials. Some materials have only a small use in ceramics, making it uneconomical to mine and produce them solely for potters. An even bigger question is what happens when supplies run out. Many raw clay supplies have already become exhausted; remember they are finite.

Plastic barrels of slop transparent glaze. In industry glaze is supplied this way as it is more cost affective and saves large amounts of energy. Photo © Kevin Millward

CHAPTER 4
COLOURS AND OXIDES

The source for colours in ceramics is a big concern for the green-minded potter. In virtually every lecture I have given on glazes and the use of colour, the first question is always about how safe an oxide is.

Lead, copper and manganese are some that raise concern. Chrome, cadmium and selenium are rarely mentioned, but these are actually more toxic, which leads me to believe that many potters are misinformed on this issue. Let us take lead as an example. Raw lead in an unfritted form has not been used since the introduction of the law banning the use of raw lead glazes and colours at the beginning of the twentieth century. Manganese is commonly quoted as being linked with the death of the studio potter, Hans Coper, but there is no proof of this.

The fact is that most oxides are perfectly safe if they are handled correctly. It is not advisable to eat while you are working. In fact, you should not eat in

Mining cobalt can be a less than pleasant process for those on the fringes of commercial mining. Photo © Getty images

the studio at all. Controlling dust in the workshop is paramount, and inhalation should be avoided by use of a mask when handling powdered materials and using dust extraction.

It is crucial for the environment that you dispose of all ceramic waste responsibly. Working in education I have many times seen a student take a large quantity of raw oxide such as cobalt, mix it with water, make two brush strokes on a small pot, and then wash the remainder down the drain. This is a despicable waste of an expensive raw material and a callous disregard for the environment. In universities this became a problem until we began charging for oxides and other costly materials, which were dispensed by the technician in appropriate amounts. It was interesting to note that when the students had to pay for these materials, the problem solved itself.

An alternative solution is to have small amounts pre-mixed in sealable containers, so you can make that one brush stroke when decorating the pot with a clean conscience.

Unfortunately, there is no organic alternative to the metal oxides we use daily in ceramics, so the best we can do is act as responsibly as possible when using them. A few potters choose not to use particulate colours on moral grounds. I have a student who chooses to use ground up blue wine bottles instead of using cobalt oxide or carbonate directly.

One of the primary ways to protect the environment is to ensure that no metal oxides enter the water course or food chain, by not disposing of oxides or glaze residue down the drain. There should be a catching system that prevents clay, glaze residue and metal oxides entering the drainage system. This should be emptied on a regular basis and the slops can be dried out on plaster bats and then disposed of responsibly. You can contact your local authority to find out where to dispose of your waste. A good idea for recycling this waste would be to make crude bricks or tiles from the waste clay dregs, fire them in the base of the kiln, and use them decoratively – for example in the garden.

Remember, the benefits of being a potter outweigh any possible danger you may think you are in if you are sensible and follow good basic housekeeping. A lifelong friend of mine, Harry Frazer, who has written many books on ceramic fault finding, once said to me that there are often more dangerous things under your kitchen sink than in your pottery.

As pottery has become increasingly popular as a hobby, more and more potters are buying pre-prepared underglaze colours in small pots, so making the life of the potter much easier and expanding decorative techniques further than just using raw oxides. Some potters have an idea of what is involved with the preparation and production of a small pot of colour, as it is just like using poster paint. Previously you would purchase colours in powder form, mix with your preferred medium (usually fat oil and turpentine), mix until no longer gritty and then paint on to your biscuit pot. An additional firing was required to burn off the oil, called hardening on, before glazing it. The great thing about pre-mixed colours is that you can glaze over them, so cutting out the extra firing.

Where do all those extra colours come from? As you know, raw oxides give a limited palette of colours, so intermixing oxides and other raw materials, then calcining them, produces a whole range of new colours. But at what cost in environmental terms?

These colours are calcined in a kiln at varying temperatures, dependent on the colour being produced. Some of the chemical reactions can be very strange; for example, chrome is green, and tin is white, but when mixed together and calcined they create pink. The recipes for making these colours are usually trade secrets and may have taken hundreds of years to develop and perfect.

I HAVE INCLUDED SOME OLD RECIPES FROM THE 1920S AS EXAMPLE (IN DRY WEIGHT)

ENAMEL BLUE

Uranium oxide 3.5 kg (7.7 lb)
Manganese oxide 1.25 kg (2.8 lb)
Refined nitre 1.5 kg (3.3 lb)
Flint glass 3.25 kg (7.2 lb)
Zinc oxide 2.25 kg (5 lb)

This would be calcined in old terms hard biscuit, i.e., about 1120°C (2048°F)

ENAMEL BROWN

Borax 7.5 kg (16.5 lb)
Ground glass 5.5 kg (12 lb)
Tin oxide 3.5 kg (7.7 lb)
Sulphur of manganese 6.5 kg (14.3 lb)
Acetate of copper 2.5 kg (5.5 lb)
Zinc oxide 1.5 kg (3.3 lb)
Refined nitre 2.5 kg (5.5 lb)
Feldspar 1.5 kg (3.3 lb)

This would be calcined in a gloss kiln, i.e., about 1060°C (1940°F).

These colours would have to be crushed, wet ground to a fine powder and dried out before use. The first stage is sometimes referred to as base colour; this is the initial calcining. From this point the required colour would be blended, then re-calcined to produce a glaze or body stain and would have no flux added. If an underglaze colour is required, you would mix the base colour then add a flux, based on such things as lead or borax. This then re-calcined to produce your underglaze colour, which is then crushed, ground and wet milled to a fine powder and dried out. At this stage it could be used in powder form or mixed with a water-based medium ready to paint on your pot. It is a similar process for onglaze colour, but more flux is added.

Gold lustre and lustres that fire in electric kilns are in general use now, used by studio and hobby potters. This is what is used to make a bright gold lustre in simplified terms; there are many variations on this method.

Take 100 g (3 oz) aqua regia (this is a mixture of nitric acid and hydrochloric acid) and use it to dissolve 10g (0.35 oz) of gold diluted with 150 ml (5.3 fl oz) of distilled water. Take 20 g (0.7 oz) potassium sulphide, dissolved in 1000 ml (35 fl oz) of distilled water. This is decomposed with 200 g (6 oz) of nitric acid. The precipitated sulphur is washed and dried. This is dissolved in 25 g (0.9 oz) of turpentine oil, 5 g (0.8 oz) of nut oil, and diluted lavender oil. The mixture is reduced until a thick syrup is achieved. Then 5 g (0.8 oz) of bismuth oxide and 1.5 g (0.05 oz) lead borate is added.

Consider how much energy is used to produce these colours. You can get some idea of the journey your pot of colour has taken, dug from the ground in a country such as Africa, shipped to Europe or America, and packaged for shipping by container to your ceramic supplier.

COLOURS AND OXIDES

Base calcination energy consumption in kWh/kg

PJK 6 Feb 2023

Base	Chemical type	Colour	Comments	Main uses	Batch weight kg	Saggars/ firing	kg/ saggar	yeild ex kiln %	kWh/kg	Calcination temp C
1	Pb, Sb	Yellow		On or in-glaze	530	82	6.4	99	4.3	960
2	Fe, Cr, Co	Black		Underglaze	515	123	4.2	96	5.3	1100
3	Pb, Sb	Orange		Glaze stain	445	69	6.4	96	6.3	1080
4	Sn, Sb	Grey		Underglaze	500	82	6.1	86	7.1	1240
5	Fe, Cr, Zn	Red Brown		On or in-glaze	270	123	2.2	95	7.8	1140
6	Cr, Sn, Ca, Pb	Mid range crimson		Underglaze	476	122	3.9	88	9.3	1240
7	Co, Si	Blue	Typical Cobalt silicate blue	Glaze stain underglaze	562	117	4.8	90	9.6	1280
8	Fe, Cr, Co	Black		Glaze stain	426	84	5.1	97	9.7	1240
9	Cr, Co	Blue green		Glaze stain underglaze	430	122	3.5	89	10.2	1240
10	Fe, Cr, Co	Black		On or in-glaze	255	121	2.1	91	10.3	1140
11	Fe, Cr	Black	Cobalt free	Underglaze	365	107	3.4	97	10.8	1260
12	Co, Al	Cobalt aluminate blue		Underglaze	429	119	3.4	72	13.5	1240
13	Cr, Sn, Ca, Pb	Strong crimson	Most complicated processing route involving two wet grinds and two calcines. However it was the strongest crimson	Glaze stain, underglaze	481	106/85 (1st then 2nd fire)	4.5/5.0	88/83	5.7/8.6	1040/1240
				Overall kWh/kg					14.1	
14	Cr, Co, Al	Blue green	Alumina hydrate	On or in-glaze	347	347	2.8	88	14.7	1240
15	Co, Al	Cobalt aluminate light blue	Alumina hydrate	Glaze stain	450	450	3.8	67	14.9	1240
16	Fe, Cr, Zn, Al	Golden brown	Alumina hydrate	On or in-glaze	212	212	1.7	86	18.0	1170
17	Cr, Sn, Ca	Pink	Lead free. Has to be calcined in two halves to achieve the heatwork	Underglaze	400	400	2.9	82	25.2	1300

Table of the energy required to produce calcined colours. This information was kindly provided by Philip Knott, who I worked with at Harrison and Mayer.

WHAT IS TOXIC?

Considering the range of raw materials and oxides available to the potter and the efforts employed to obtain them, just how dangerous are they? We can break them down into two types: toxic and non-toxic. The non-toxic materials still have the potential to do harm to the potter due to the inhalation of dusty materials that have a high silica content; in other words, most of the glaze and clays we use. Dust, as any potter knows, is never a good thing in the studio or workshop and every method of keeping the dust down should be employed. Not sweeping up, not dusting down surfaces, and wearing suitable overalls to prevent the accumulation of dry clay on surfaces and the floor are all good practice. Fettling of dry clay should not be carried out without the appropriate extraction and great care should be taken when mixing dry ingredients for glazes. Always put dry materials into water, never the other way around.

Galena mineral before processing into its usable but toxic form. Photo © Getty images

The weighing and mixing of dry materials should always be done under extraction if possible. If not, use a wet cloth over the bucket and wear an appropriate mask. Always mop up spillages immediately. Never use a vacuum cleaner unless it has a hepa filter or is a wet & dry cleaner. Silica is linked with silicosis, an accumulation of particle inhalation which can seriously affect the efficiency of your lungs. This damage is permanent and there is no effective treatment. The main source of silica dust in the studio is usually clay. Raw materials are mainly supplied in a dry form silica. Quartz and flint are pure forms of silica, so great care should be taken with these. You may have noticed that flint is supplied slightly damp to prevent the raising of dust. I often see hobby potters sanding their biscuit ware to remove faults. This is not a good idea as you are creating even more dust and in addition, the dust on your pots can cause crawling of the glaze. Cleaning the dust off with a wet sponge is also not advisable as it can encourage the accumulation of soluble salts on the surface of your biscuit ware, affecting the glaze adhesion and resulting in shivering. If your pots are dusty, you should brush the ware with a stiff natural hairbrush under extraction, wearing a good mask.

METAL OXIDES

All oxides should be treated with the greatest of respect as heavy metals are toxic when inhaled or ingested. Some are more toxic than others and their effect is cumulative and potentially carcinogenic. The substance people usually think of first is lead. The use of raw lead oxide is prohibited, and exposure to it is usually in the form of a glaze frit such as lead bisilicate. In this form it is considered safe, as the lead is fixed with the silica, but its dust still represents a danger. The outlawing of raw lead glazes in the twentieth century was an acknowledgment to the potter of its toxicity. In the days of raw lead glazes, the ware boards used were tipped with red paint to indicate they were contaminated.

It is worth looking at the oxides we commonly use in our day-to-day work that represent a possible hazard to our health, but I do stress that providing you use them as directed they are perfectly safe. I have noticed with the rise in popularity of hobby pottery that food and drink are being consumed whilst working with ceramic materials and as previously mentioned this is definitely bad practice.

Here is a basic breakdown of which oxides should be treated with respect and why:

IRON OXIDE
Not normally regarded as toxic, inhalation and ingestion should be avoided; can also cause discolouration of the skin, as anyone who has dipped tenmoku glazes by hand will testify.

CHROME
A potential carcinogen and any form of inhalation or ingestion should be avoided.

COBALT
Inhalation and ingestion should be avoided; can cause allergies and irritation.

COPPER
Rarely involved in poisoning and is present in many domestic items. Copper sulphate on the other hand is toxic and should be avoided in ceramics.

LITHIUM CARBONATE
Partially soluble and when ingested may cause severe lesions in the bone marrow and symptoms similar to pernicious anaemia and leukaemia. Not usually fatal.

MANGANESE DIOXIDE
Can be dangerous if ingested or inhaled. Can permanently affect the central nervous system.

CADMIUM OXIDE
Very toxic. Inhalation or ingestion should be avoided.

SELENIUM
Very toxic. Inhalation and ingestion should be avoided. Most potters do not come into contact with cadmium or selenium in its raw state, only in pre-prepared glazes, underglaze colours, onglaze colours and stains.

VANADIUM PENTOXIDE
Can cause severe irritation if inhaled.

ZINC OXIDE
Inhalation and ingestion can cause temporary illness.

TIN OXIDE
Avoid inhalation as it can cause pigmentation of the lungs, coughing and shortness of breath.

URANIUM OXIDE
Gives great yellows. Not used today as it is radioactive.

RESPONSIBLE PREPARATION, STORAGE AND RECLAIMING

TIP: Water is a good example of something that is easy to recycle. Put all your wet slops from throwing, glazing and general cleaning into a large tub. Let the solids settle out so you can pour off the water. You can then use it in the garden or simply pour it down the drain with a clean conscience. Dispose of the solids responsibly.

LEGAL & ETHICAL DISPOSAL

The disposal of ceramic materials – plaster, clay, colours and metal oxides – will very much depend on the country you are in and the requirements of your local authority. It is inadvisable to put ceramic waste in your domestic bin to prevent heavy metals getting into the water course. Many local authorities now have comprehensive local waste management facilities, but you may have to pay for this service. Find out where you can dispose of your waste safely and if in doubt maybe save it up, mix with scrap clay and fire it, thus rendering it much safer.

TIP: Remember there are probably more dangerous things under your sink!

Copper carbonate as the potter obtains it in a nice plastic pot. In the past it was supplied in a paper bag. Photo © Kevin Millward

CHAPTER 5

FUEL FOR THOUGHT

For many potters the area of most concern is the energy consumed when firing the kiln. The electric kiln is probably the most viable way of being green, as the power station can burn fuel far more efficiently than you could burning gas or oil in a kiln. If only electricity produced from renewables could be used, then we could be carbon neutral.

What are our options if we choose to do reduction firing in a gas, wood or oil-fired kiln? Other fuels still have an impact as the methods of production, while cleaner than their existing counterparts, still have a carbon footprint. Most people have heard of the environmental advantages of hydrogen, as its only by-product when burned is water. But it may not be commonly known that hydrogen is usually produced from fossil fuels. So it is not as green as people imagine! If hydrogen could be made from renewables that would be fantastic, but unfortunately that is not the case at the moment. Currently, the UK is exploring the

Wood stacked for drying after cutting to the correct size and shape for the fire box. Photo © Kevin Millward

CERAMICS: A GREEN APPROACH

Elements inside electric reduction firing kiln in Japan. Photo © Kevin Millward

possibility of putting up to 20% hydrogen into natural gas in order to extend our dwindling supplies.

In Japan, scientists may have found a way of producing low cost, clean hydrogen. So how was this developed? A new type of nuclear reactor that is gas cooled, rather than water cooled, enables the reactor to run at a much higher temperature. Lessons were learned the hard way with the meltdown at Fukushima and now it is more or less impossible for the new reactor to go into meltdown. The higher operating temperature enables the waste heat to be used to produce so-called red hydrogen, cleanly and at a competitive price. As the risk of meltdown is low and such vast quantities of water are not required, the reactor can be positioned closer to industrial centres, reducing the costly distribution of the hydrogen. The downside is that there is still nuclear waste needing to be stored for thousands of years.

It seems to come down to which is the lesser of two evils: fossil fuels or nuclear generation, until renewables like wind, water or solar power become more widely available for use in the national grid. The initial financial investment of installation costs, and the many years taken before the financial benefit is felt, deters many potters, but this does not mean it isn't possible.

TIP: Have you considered using tariffs like Economy 7 and firing overnight to take advantage of cheaper electricity? In addition to cheaper fuel, you have the bonus of having a warm studio during the day. As most deals don't start until after midnight, you may have to pre-heat on the higher tariff before the lower one begins. Be aware that some suppliers do have a penalty for use outside the agreed times.

PRODUCING YOUR OWN GAS

You may wish to consider becoming more self-sufficient by producing your own biogas with a waste food digester. This can be a viable option if you have enough food waste and only want to run a cooker or central heating system, but what are the chances of powering even a small kiln with this method?

Your first challenge is that most small domestic digesters won't supply enough gas on a continuous flow to fire even a small kiln. Next you would need a method of compressing the gas and suitable storage for the volumes of gas required to fire your kiln. It is possible to do all this, but it requires you to build a digester that will produce a decent quantity of gas on an ongoing basis and enough waste food or other organic waste to feed it.

If this was something you might consider, ensure you research the legality of pursuing the project, depending on where you are in the world and your local authority. Be realistic about your competency to undertake the work, and if in any doubt consult an expert propane gas fitter to undertake the work for you. Make sure everything is legal and in place to produce the gas and that you have a way of storing up to, say, 1000 litres. The next step is to compress this into propane cylinders, which is not that simple. It is unsafe to transfer compressed gas from your digester directly into the steel cylinders, as the moisture content is too high and will promote rust, which will degrade your cylinders. The gas will have to be filtered to remove the water. There are many examples of this process showing simply-constructed digesters and filters built from readily-available components off the shelf, as this first stage is low pressure. The compression is achieved by adapting a compressor used for paint spraying but using standard propane fittings to compress the gas into the propane cylinder, not using the air storage tank. You must make sure any air is kept out of the cylinder or vent it out before storing, as this can be explosive.

YouTube is a fantastic resource for information on building this system. To counter the argument that this process is still producing pollution and CO_2, methane is less polluting when burnt and it is a major greenhouse gas when expelled into the atmosphere. CO_2 makes up about 0.04% of the atmosphere; anything less than 0.02% endangers plant life. Methane makes up about 0.00017% and is 80 times more harmful than CO_2.

Another source of fuel for your kiln is waste cooking oil, a waste product of fast-food shops. At one time they would have thanked you to take it away for them, but now, as the base for biofuel, it no longer costs them money to have it removed. However, a small independent business may be glad for you to remove it. To prepare the oil for burning simply filter out any food debris. Some oils, however, require heating to become a liquid again; they can become solid at room temperature and not flow through the pipework to the burner or fire box. Some potters will mix it with paraffin to make it more fluid.

PREPARATION OF WASTE OIL FOR BURNING

You may have considered using waste engine oil, either synthetic or petroleum based, but I do not recommend burning this type of waste oil due to the levels of contamination and pollution it would release. In many countries it is also prohibited. There are well established methods for the decontamination and preparation of this type of oil, but it may not be cost effective for most potters to entertain.

UNCONTAMINATED WASTE OIL AND SAWDUST

I would not advocate burning waste oil and sawdust in urban areas, but it could be used to supplement a wood-firing kiln in a more rural area. Fine sawdust is not normally used for animal bedding as it is too fine and can cause respiratory problems, so is more readily available to use as a potential fuel. It is not recommended to put sawdust directly into a fire box as it can spontaneously combust; mixing it with oil can make a usable fuel to help supplement a wood firing. Horse dung, straw and wood shavings, when dried, compressed and perhaps put through a pug mill, can also be used as an alternative fuel source. A problem with this type of fuel is being able to dry it out without expending too much energy. Animal dung can contain up to 80% water and long-term storage can compromise its calorific value, as

Waste cooking oil stored in large plastic barrels before it is sent off for prepossessing into bio diesel.
Photo © Kevin Millward

when mixed with wood shavings and straw it has approximately the same energy density as wood. The ash residue from this fuel has a high melting point, as its grass content will introduce silica to the mix. You would have to consider how this would flux on the clay bodies being fired. It might depend on whether you were using this as fuel only and any flashing is a bonus, or if you are intending to draw the ash through the chamber to glaze the pots.

To give you some idea of the quantity of raw materials available to potentially create this fuel, 800 million metric tonnes is produced annually. It is interesting to note that the chlorine levels produced are twice as high compared to wood. This source of fuel would not suit everyone, as the smell can be off-putting for some. I once built a small Acoma kiln for a television program that was fired on dried cow dung and a student of mine, whose family were farmers, collected and dried out the cow pats that were needed to fire the kiln. It was quite a task. Animal dung from grass-fed animals is a staple fuel in many third world countries, and it could be argued that it is carbon neutral as it's mainly grass. If you have an ongoing supply of dung, it may be more efficient to use it in a bio digester.

Many potters look to Japan for inspiration in their ceramics, a country steeped in long-held traditions of form, function and ideology, including the romance of large wood-fired kilns and the results they deliver. However, it may come as a surprise to find that due to very strict controls regarding polluting processes, the ability of Japanese potters to fire on wood or even gas-fired reduction kilns can be problematic. For many years, well-established potters with enviable reputations have been reduction firing in electric kilns. To many potters this would be, at the very least, not sound practice, even heresy, due to the perceived detrimental effects on the kiln and elements, but this is not necessarily the case. I have recounted to students on many occasions that when I was at art school in the late sixties and early seventies, under the instruction of our tutor, we would do reduction firings in the front-loading electric kilns, even though we had a gas kiln. At this time all the kilns were operated manually by Sunvic energy regulators that controlled the amount of energy distributed to the elements.

The kilns had thermocouples and an analogue pyrometer, with pyrometric cones to determine when the firing was completed. As with a gas, oil or wood-fired kiln, the point at which you begin the reduction is very much the decision of the potter. I have normally started reduction at about 1060°C (1940°F), unless it's a carbon trap shino, when it would be about 850°C (1562°F). The method is to open the top damper with extraction on over the kiln or to make sure the room is well ventilated but be aware of smoke detectors.

My tutor seemed to prefer moth balls to create reduction, although I don't know how safe this was – but it was the seventies! My preference was small slivers of dry wood which were delivered through the front bung hole. The kiln doesn't need a constant delivery of wood, only once the previous delivery has burnt away. If you wish to have only a glaze reduction as opposed to body and glaze reduction, you can start much later at say 1200°C (2192°F), which obviously has less wear on your elements.

For potters who may not understand how this technique impacts the elements, the elements are made of different gauges of Kanthal wire that are dependent on the size of kiln and the amount of power available; roughly speaking the more power you have (for example three-phase) the thicker the element are and therefore more reduction firings can be done. When the elements are new, you will notice they have a metallic shine, but after the first firing they take on a matt, grey appearance, which is the oxidisation of their protective coating. This coating is stripped during each reduction firing, removing a small quantity of the wire's surface, lowering its resistance and shortening its life span. It is recommended that an oxidised firing is done between reduction firings. Limiting the number of reduction firings will safeguard the life of the elements.

In industry, silicon carbide elements are used, which are not affected by the reduction atmosphere, but they are fragile and expensive. In Japan they have found methods to avoid the problem of element degradation. They use different types of elements that are much thicker and more like rods or strips than wire. This increased thickness helps to maintain the resistance for a greater period of time, or the elements can be shielded in porcelain tubes to protect them from the reduction atmosphere. Interestingly, this method was used in kilns fired on coal gas, to protect the pots from the sulphur in town gas during the changeover from coal-fired bottle ovens.

The methods for inducing reduction vary. Some use a small propane gas burner that provides a stream of gas into the chamber, thereby consuming the oxygen to create a controlled state of reduction. Another method uses a small chamber

Reduction firing in a top loader, feeding in small slivers of wood. My old tutor liked to use moth balls. Photo © Kevin Millward

charged with charcoal or wood chips, having the advantage that the firing can be controlled by the operator, although this does not mean you can leave the kiln unattended. Even these methods can be problematic as far as emissions are concerned and is probably subject to the local authority's green incentives.

I'm puzzled as to why this type of kiln has not made inroads into the American and European markets. I have discussed this with kiln manufacturers who feel there is no real demand for it. Maybe, as with so many things, improvements in emissions will not occur until they are forced by legislation. I have sometimes found that potters are suspicious of technology encroaching on their craft and even if a more efficient alternative is available, they often prefer to stick with what they know, even if it more costly to them and the environment.

I have always thought that the quality of a finished pot is more important than whether it has been fired in a wood-fuelled kiln or electric kiln. No firing process will make a bad pot into a good one. A good pot is a good pot, irrespective of how it is made or fired. As an example, some of Japan's leading potters prefer to fire their celadon in electric kilns and create the reduction atmosphere when the glaze is forming, giving a superior effect.

At Clay College, some students and I built several variations on our experimental gas and wood-fired kilns. The lack of smoke and smell associated with the firings with wood was noticeable, along with the fact we still got excellent reduction. I explained it would be easy to get it to smoke, but, although it may look spectacular, it would not be achieving much other than wasting fuel. I altered the damper on the flue and the kiln smoked, demonstrating the wasted energy, as smoke poured from every gap in the kiln bricks. The smoke and smell of reduction, although spectacular and exciting, is a waste of fuel. Once I adjusted the damper, the draw on the kiln pulled the smoke back in. It was interesting to note that the block of studios and offices above and adjacent to the kiln only complained once, and surprisingly it was not about the smell of smoke but that of the rubber solution used as a resist. From that point on we no longer used rubber solution, only wax emulations, and there were no more complaints.

CHAPTER 6

KILNS AND POWER REQUIREMENTS

One area where potters can reduce their firing costs is in the density and amount of kiln furniture used, the thickness of the kiln shelves, and the types of props used.

The ceramics industry is aware of these issues and some larger kiln shelves have a hollow section like corrugated cardboard, as unnecessary density means higher fuel costs, probably their biggest overhead. Potters are often unaware that most kiln shelves are manufactured solely for industry, and their specifications are for the temperatures, firing cycles and atmospheres used in the production of industrial ceramic products, including tableware, sanitary ware and high-tech specialist products. They select the appropriate product for their operating temperature and its total density, which is directly related to the amount of fuel needed, as the fuel is required to fire the pots, not heat the kiln furniture.

Silicon carbide kiln shelves in use in reduction gas kiln.
Photo © John Jelfs

Much thought goes into the design of the kiln furniture to give function, durability and ease of handling with minimum density. Many of the firing processes we use for reduction, the temperatures reached, and the amount of heat work involved, have a detrimental effect on the kiln furniture which is not designed to withstand this type of firing. Overly thick kiln shelves and solid props, often made from sawn up fire bricks, mean that large quantities of fuel are consumed in just heating up the kiln furniture. It is also important to choose the correct insulating fire brick. Consider something as simple as not building a large bag wall out of fire brick, when an insulating fire brick would suffice. Or if you must use fire bricks, think about using the half thickness bricks. Remember, the more density put into the kiln chamber, the more fuel you waste. Speak to your local supplier of kiln furniture and ask for their advice on the best types to go for, relative to your firing process and the type of product you are firing, as there may be a more efficient system to use. Try using foot placing cranks for firing plates to dramatically reduce the amount of density in your kiln. Many potters may concentrate on trying to keep their costs down when setting up their kiln furniture, even if this means constantly wasting fuel, rather than spending up front and saving money over a longer period.

The majority of refractories are made for industry, not for studio potters, and therefore are only supplied in bulk. This has created a gap in the market for small companies to wholesale and customise kiln furniture for the craft sector. Many forms of kiln furniture are made for a specific product, for example cranks for plates, but your supplier is always up for a discussion on appropriate ways to customise existing products to your requirements. Again, the correct kiln furniture can have a dramatic effect on the efficiency and packing density of your kiln, so talk to the experts to get their advice on what you need. This could be as simple as whether your kiln shelves are thicker than they need to be, relative to the temperature you are firing to. If you are firing to very high temperatures and are about to build a new kiln, it may be prudent to talk to your supplier first to see what is available from stock and work your build around this, instead of building something and finding you have to spend far more on having them cut to size.

Hollow section kiln shelves used to reduce thermal mass in the kiln, not suitable for aggressive firing processes.
Photo © Kevin Millward

JOHN JELF STORY
SILICON CARBIDE SHELVES

John Jelf, who has a lifetime's experience of making and firing pottery, became very aware of the rising cost of fuel and its impact on his profit margins. It can be difficult for small producers to pass on rising material and overhead costs to the consumer, whether this be an individual or shops and galleries. Ever aware that it is unlikely that fuel costs will return to their previous levels, John looked at other areas that could enable him to reduce his fuel costs. As we have discussed, much fuel is consumed by repeatedly

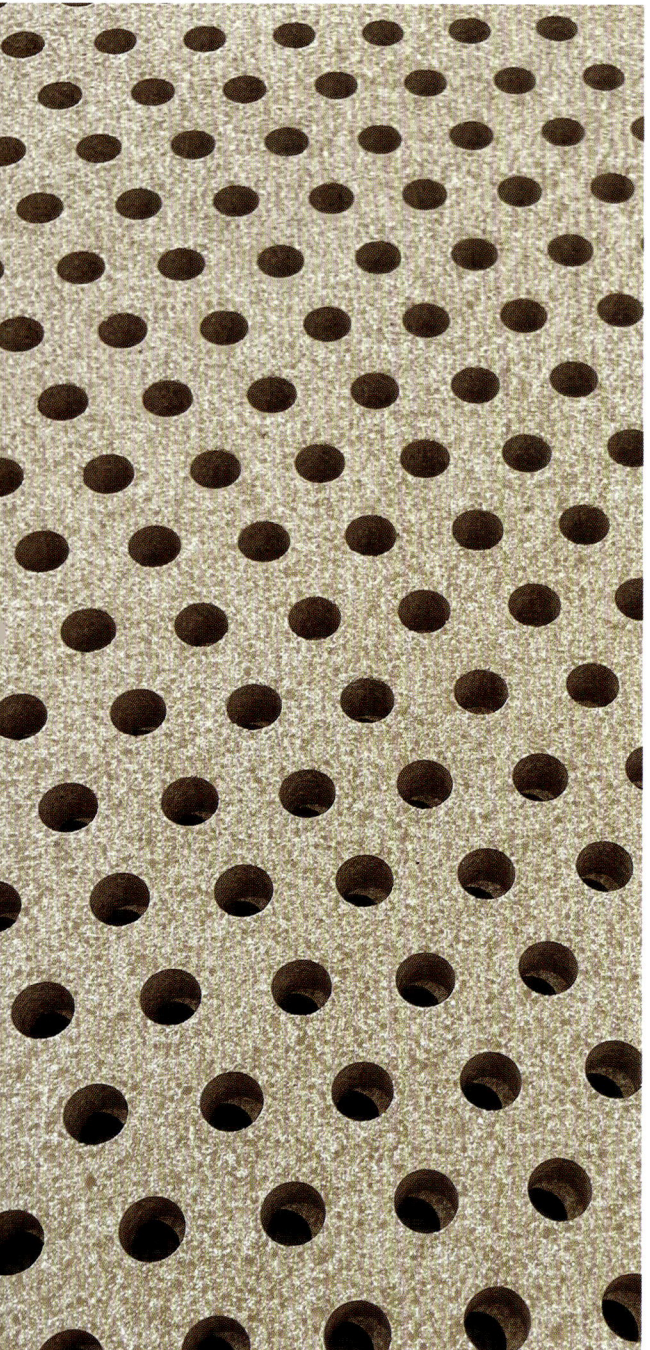

Perforated kiln shelves to reduce thermal mass, again not suitable for all firing methods.
Photo © Kevin Millward

heating the refractories. One simple way of reducing this is to change over to silicon carbide shelves, as they have the advantage of being stronger, thinner, and less likely to warp. Due to their reduced thermal mass, large savings can be made on fuel consumption.

If it's that simple why isn't every potter doing it? The upfront costs can be prohibitive, and some potters are not willing or able to wait for the investment to pay off. John had to wait about a year to appreciate his savings, which were considerable, in the region of one third. He did however find the shelves were not suitable for his soda firings as it caused the glaze to drip off onto the pots below. With the high cost of fuel, our thinking will have to change. We have all seen images of self-built potters' kilns with hard fire bricks cut up to make props that are solid; these require large amounts of energy just to heat up but are also capable of retaining the heat. This can cause cracking of the kiln shelf in the corners as it creates heat retention. This is why props are hollow; it requires less energy to heat them up and they also lose the heat faster. If you don't want to buy kiln props, they are simple to make if you have an extruder, wad box or pug mill. I have included some recipes for refractory bodies if you wish to have a go. Most kiln props are made in long lengths and fired to the maturing temperature for that body, and then they are cut down to size with a diamond saw. I have used a small DIY tile cutter to cut props down very accurately to the same length, a very safe and cheap task as the cutting is wet, so dust free. The type of HTI (high temperature insulator) used can have dramatic effects on fuel consumption and the rate at which the kiln temperature rises and cools. Have you ever marvelled at how thin the brick walls are on your top loading kiln? They are designed to allow the temperature to both rise and cool quickly. If the rise in temperature is too slow, as in the case of tired or failed elements, you will notice that the external casing gets very hot and does not reach temperature. This is because once the heat penetrates through the brick, its efficiency is lost. It is why larger kilns have thicker walls, providing greater insulation but at a higher cost. Most kilns for the hobby and studio sector today are built using HTI 23 as they offer the best efficiency against the cost of build and firing but unfortunately are much more fragile that other bricks.

HTI 23

A silica alumina brick ideal for fuel efficiency, lower cost, and easy to cut for arches with a brick saw. They are very fragile due to the abundance of air pockets that provide the insulation, therefore a bad conductor of heat. The china clay they are made from provides structure.

HTI 26

Made from the same basic materials but with fewer air pockets. They are not as fuel efficient but much stronger and are traditionally the favourite type of brick for the reduction firing potter building his own kiln. They have been replaced mainly by HTI 23.

HTI 28

Again, made from the same basic materials with the least amount of air pockets, reducing the efficiency further. They are much stronger and have a smoother face which is better suited for use in wood, salt and soda kilns as they are more resistant to the corrosive atmosphere in the kiln.

TIP: You can make your own HTIs. There are many recipes available, but they are basically china clay and sawdust.

Modern bricks don't use sawdust due to pollution. The drawback is if you want to build a kiln out of your own bricks, how do you fire them? Some potters build the kiln with raw unfired brick and fire them in situ. It's worth noting that on firing they create a lot of smoke and shrink a lot. This can be helped by adding molochite to the mix.

HEAVES, FIRE BRICK

The quality, price and efficiency of this brick can vary dramatically. They can be cheaper than the HTIs and you will use much more fuel, but they will last longer. Imagine the scenario of building a gas kiln using the most efficient type of brick, the correctly designed forced air burners and silicon carbide shelves, combined with the correct firing program. The savings would be dramatic, both in the terms of fuel but also the environment.

The efficiency of your refractories can be improved and protected from aggressive firing processes and your fuel costs can be cut by the use of furnace coatings. These can be purchased, or you can make your own from ingredients such as zirconium silicate, calcined alumina, china clay and possible silicon carbide.

Be warned though, they can cause degrading of the brick face, known as spalling. There are also coatings that can be used to protect your elements. One area that is gaining more interest, from potters using electric kilns in particular, is the formulation or purchase of glazes that mature at a lower temperature.

Using a thermocouple and pyrometer is not always an accurate indication of heat work, and using cones will give you a more accurate firing; those who use cones may refer to a cone 10 or 11 when firing stoneware at 1280°C (2336°F). The amount of energy required to attain these high temperatures is considerable, let alone the wear and tear on the kiln and elements, which has encouraged many potters to reduce the temperature or cone number down to perhaps 1240°C (2300°F) or cone 6. There is a good book on this specialised area called *Cone 6* by Michael Bailey. By reducing the temperature, the work is not compromised but there are savings on all fronts; fuel, time and the longevity of your kiln. You could also consider whether the firing cycle is too long, to avoid using more fuel than is necessary. To attain the higher cone numbers if you are unable to reformulate your glazes, remember that it is heat work, not temperature, that counts, so a more rapid rise to the top temperature, combined with a soak until your cone goes down, will result in the desired glaze finish while using less fuel.

One area that potters don't often consider is the amount of heat given out by their kilns after the firing is complete. In many cases, other than warming the studio, this is wasted heat. What possible ways might there be of saving this energy or redirecting it to other parts of your studio or home?

Unfortunately, when your kilns are firing on the way up to temperature, venting this heat to other areas that are being used is not advisable because of possible toxic fumes, but it is possible to move this heat on the cooling cycle. In the past I have used fans pointed at the kiln's external casing to distribute the warm air to the rest of the studio. One possibility, though a bit more complicated, is to create some form of heat exchanger. This may need the assistance of a plumber if it's an area of DIY that is outside your comfort zone. This could be as simple as a cold-water radiator system in your studio, using heat from the kiln to warm the water, then pumping it around the other radiators in your studio or home. For safety this should be an open system in the unlikely event of the water boiling. A closed system should have a pressure release valve incorporated. If you want heat exchangers to be incorporated into your central heating system, then I would recommend consulting a qualified heating engineer. There is such a system commercially available, but it can be expensive, and the cost may be too high and override any long-term benefits. I have often wondered why the ceramics industry never developed a way of utilising its waste heat.

Things to consider that can create fumes when firing ceramics:

Clays with a high carbonaceous content, i.e., some ball clays
Fat oil
Pure turpentine
Wax emulsion
Candle wax
Latex solution
Decorating media for colour
Carboxymethylcellulose (CMC)
Lustres
Gum Arabic
Gum tragacanth

CHAPTER 7

ATMOSPHERES

The process of firing your pots can have a dramatic effect on how the clay body and glaze appear. Probably the most common method of firing, irrespective of the fuel used, is oxidisation, in both industry and the hobby and studio sector. This process ensures that there is enough oxygen in the kiln for a clean firing with consistent results.

In some ways this is the holy grail for potters, who use it as a way of improving the qualities of the glaze body and decoration. Until the abolition of coal-fired kilns, the only way they could ensure a clean firing was to place the pots in saggars for protection, but today these are no longer needed as modern electric and gas kilns produce a consistent, clean, oxidising atmosphere. Paradoxically, now that we are able to consistently achieve a clean firing kiln, we want to fire pots in dirty atmospheres for their desirable effects. A prime example of this is during reduction firing where the amount of oxygen into the kiln is restricted. Essentially, the kiln is forced into a state where the fuel cannot combust efficiently and for it to be able to do this, oxygen is taken from both

Saggars made in China transported for use at Hamadas kiln in Japan. Photo © Kevin Millward

the glaze and body, creating a range of colours – particularly in the glaze – that would not be available otherwise.

What does this mean for the environment? This process burns fuel inefficiently and releases more carbon monoxide into the environment. I can personally testify to its dangers. When firing in reduction you must ensure there is good ventilation or extraction available, especially when the kiln is indoors. Reduction can be achieved at a wide range of temperatures and is not just restricted to stoneware. It can also be induced by the addition of fine silicon carbide into a glaze of less than 3%. And it can be achieved outside the kiln, as in the case of raku. It is worth knowing that the post reduction in raku is a modern incarnation and did not form part of the traditional Japanese technique.

Reduction is also used in traditional lustre firing at low temperatures of approximately 800°C (1472°F). However, commercial lustres are formulated with an ingredient to create their own localised reduction atmosphere in electric kilns, to achieve a similar effect. It should be noted that the fumes from the commercial lustres, when applied and fired, are extremely toxic and excellent ventilation is required.

The interest of the studio potter in all things reduction has led to the adoption of techniques usually associated with industry, such as salt firing, a technique developed in Germany before finding its way to the UK. This practice involves salt being introduced into the kiln, wet or dry, which will volatilise before combining with the silica in the body to form the glaze surface. The method was rapidly adopted by potters as a quick way of producing functional domestic items, as the ware was normally only fired once. Its strength and resistance to acid and alkaline attack made it ideal for sanitary wares and drainpipes, but the process was outlawed in the mid-twentieth century due to the release of sulphur dioxide, creating pollution and acid rain. The toxic gases present around the kiln concerned many potters, leading to a move away from the use of salt to that of soda. One way this can introduced is in a solution with water which is then sprayed into the kiln at around 1240°C (2264°F), but this is not the only way. The reaction on the surface of the pot that occurs with soda is more directional, whereas salt fills the whole kiln chamber for a

Salt glazed pots by student at Clay College fired on natural gas forced air burners. Photo © Kevin Millward

more overall effect. Salt and soda glazes are both clear and the colour comes from the clay body or slips applied to the surface at the clay or biscuit stage. A small number of potters will combine processes such as wood firing, salt or soda glazing alongside reduction to achieve the effects desired.

How does this all stack up as far as the environment is concerned? Potters firing in this way could argue that the quality of the pots produced and the longevity of the articles outweigh the relatively small amounts of pollution produced, especially when compared to industry. Should this be for the conscience of the individual potter or controlled by legislation?

LEAD RELEASE

Can a good clear glaze on terracotta be attained without lead and how much lead is safe? Officially, there is no such thing as a lead-free glaze, due to natural lead in the environment; the phrase used instead is "no lead is added in manufacturing". The fear associated with lead glazes is, in most instances,

Red earthenware glaze tests showing the qualities of different additions of lead in the glaze. Photo © Kevin Millward

exaggerated since the introduction of fritting. This is where the lead is fused with silica to form a glass, locking the lead into the glaze, and it cannot be ingested or absorbed during application due to this manufacturing process. However, lead can leach from the finished pot in certain circumstances, where the glaze surface is subject to acid for extended periods of time. The test for lead release, normally applied to functional wares intended for food and drink, varies from country to country and is usually a necessary safety requirement for export. The test is normally split into flatware (plates and dishes) and hollowware (mugs, cups, bowls and jugs). A formula is used to calculate what is considered flat or hollow, the latter having the potential to be used for storing acidic liquids for extended periods of time. For example, storing orange juice in the fridge in a jug. This situation would constitute a high risk of lead being leached into the juice and could be greatly exacerbated by the presence of copper in the glaze, a fact we should all be aware of. Copper and lead glazes should only be used for decorative purposes and in some states in the USA, lead-glazed domestic ware is not allowed at all. There is a strange anomaly with lead glazes, in that higher amounts of lead in the glaze can actually produce a safer glaze when tested for lead release.

SAGGAR FIRING

Historically, the saggar's function was to protect the pots from the products of combustion. Today we can reverse this process by placing combustible materials such as seaweed or wood in the saggar, then firing in an electric or gas kiln to produce the same effects as a dirty firing, whilst protecting the bricks that would normally only withstand a standard reduction firing. Another way to achieve this is by positioning the combustibles through a hole in the saggar corresponding with the bung hole of a front-loading kiln door or lid of a top loader. A further vent could be created, aligned with the top damper present on most front loaders. In all cases, adequate ventilation must be provided. This technique can enable wood, salt and reduction firing in an electric kiln without affecting the elements. This system works just as well in a kiln fired on oil or gas made from, for example, HTI 23 bricks, and will return efficient firing with lower use of fuel. The term for this type of kiln is a muffle oven.

CHAPTER 8

FIRING

Many kiln manufacturers are abandoning the construction of gas kilns for more efficient and cost-effective electric kilns. There is an insufficient volume of natural gas with enough pressure for kilns in a home studio setting. Furthermore, there are health and safety issues associated with the bulk storage of propane cylinders, due to the almost impossible legislation in the UK, regarding the space required to store these tanks, which must also be caged when not in use. This has consequently reduced the demand for gas kilns by small scale producers and hobbyists and made it no longer financially viable for manufacturers to carry on production. More advanced potters may still build their own gas kilns but will face the same problems with both supply and storage of the gas. In most cases, small scale production potters simply cannot afford the upfront cost of installing an industrial mains natural gas supply.

I well remember the instillation of the mains gas supply to my workshop in my hometown of Leek. This involved a team of men who, over a few days, dug a hole in the main street, a deep trench down the alley at the

Kiln fired on propane. This kiln is commercially manufactured to be as efficient as possible. Photo © Adam Frew

back of my studio and then cut a hole in the studio wall to install the industrial gas meter. They then had to back fill the trench and hole in the road and replace the tarmac. As the studio was just under the maximum distance from the mains in the street, it was a fixed fee of £150, which in the mid-seventies was more than a good week's wage. British Gas was state owned at this time and was promoting the use of gas as both a domestic and industrial fuel, hence the relatively low installation cost. Today this would cost thousands of pounds to install and is therefore driving potters to use other fuels.

Precious metals used in lustres and onglaze colours have a great impact on the environment. However, their extraction, preparation and manufacturing into a usable form is not our only concern. The mediums used to enable the application of lustres and onglaze colours are usually based on natural oils like turpentine and fat oil (evaporated turpentine), which act as a carrier to the ceramic surface, holding it in position until dry and so avoiding smudges when handled. The medium burns away during firing before the colour fluxes and fixes to the glaze surface, creating toxic fumes in the process. When printing any transfers, both the cover coat and medium has also to be burnt away. During a lustre

Propane bottles interlinked for use on modern gas kiln in Japan. Photo © Kevin Millward

ware firing, the medium holding the lustre creates a localised reduction atmosphere when burnt away, again creating toxic fumes. Experiments have taken place to produce transfers using water-based mediums that are more pleasant to handle in the printing stage and do not give off toxic fumes when fired, but these are not compatible with a process that requires soaking in water.

These processes discharge noxious emissions from the kiln that are normally vented to the atmosphere. I remember well the smell in the streets of Stoke-on-Trent when the enamel kilns were firing, only surpassed by the noxious smell produced by the oil and gum Arabic burning off from the kilns firing bone china flowers. Gum Arabic is added to clay to make it more plastic to model and give it strength when dry prior to firing; olive oil is used as a lubricant, as water would make the clay too sticky.

How do we move forward in a responsible way that lowers our impact on the environment, without preventing our growth and creativity as potters? A simple but very effective way to dramatically reducing the waste of materials and fuel is to only fire your best work. Don't fire pots you are not happy with. Look long and hard at your work and be more selective at the clay stage, as up until you fire your work it can be recycled. We are all guilty of this reluctance to discard work, but hobbyists are probably the worst offenders as they usually love everything they make. Step back for a moment at the clay stage and ask yourself; "Should I recycle this and make

A box of Bullers rings as used in industry. Photo © Kevin Millward

a better one?" You don't have to keep everything! Maybe the exception to this are the pots made both by adults and children exploring their first contact with clay; nobody wants to spoil the joy of that first pot, taken home with pride to excited parents or family.

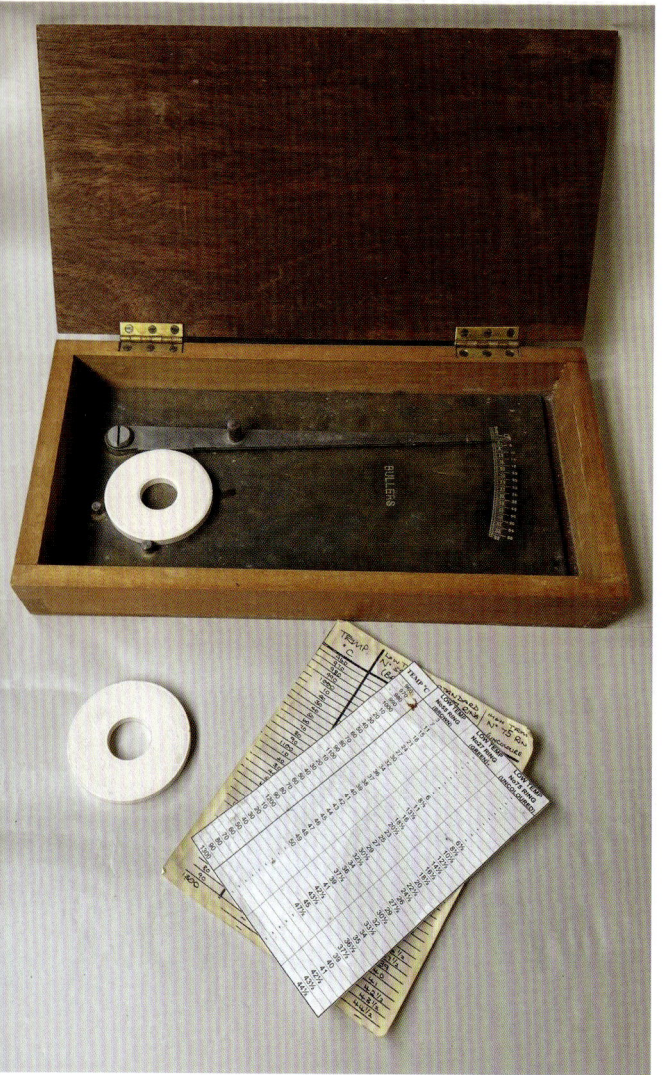

My old brass gauge for reading Bullers rings, a low-tech solution for a very precise requirement still used to this day. Photo © Kevin Millward

Both professional potters and those in education are not exempt from this. I find students, given the chance, will fire most of what they make even if it's only to biscuit. We have all seen it; biscuit ware sitting on shelves for years only to end up in a skip. Consider the financial and environmental cost of this wastage. If it's no good, you don't need it. Don't fire it, recycle it instead. If you have a process that has a high loss factor, maybe you need to take a step back to see if you can improve the process and identify any problem areas where you could reduce this. If you are losing more than 10% of what you produce, you have cause for concern. Interestingly, industry accepts a greater loss rate on flatware, i.e., plates, saucers etc. due to the complexities in their production.

If we exist in a world that won't accept some of the processes potters use to create their identity – for example wood, salt and soda firings – we need to discover ways to overcome these constraints with some lateral thinking, allowing us to create pots with unique qualities that don't fall into bland uniformity. Many studio potters vilify industrial potters for the lack of individuality; the uniformity of what they produce is, some may say, "safe". This is very often due to companies not being willing to take risks on new shapes, decoration or fashionable trends that by the time they get into production may have moved on. Very often the ceramic industry plays it safe, as the complexities of production may lead to significant losses that cannot be overcome, making it uncompetitive in the marketplace. For many years, UK-based ceramic manufacturers have perfected their production methods of earthenware and bone china shapes and forms, allowing for low loss factors, up to the glazed white ware stage which can then be decorated with onglaze transfers and gold lustre. This gives the studio potter a great advantage over industry in that we can adapt, create and be flexible to upcoming trends in the market. I often say to my students that it doesn't matter how obscure your idea is, there will be more than one person who will like it as much as you do. You can be an innovator!

Where we can learn from industry is to use technology to make us more efficient and less wasteful, as being in control of your process does not compromise your creativity but enhances it. There's no getting away from the fact that making ceramics is a combination of creativity, understanding the materials and having the technical knowledge to see it through the different stages required to produce a complete, viable piece of ceramic. If these steps are disregarded, it may lead to losses and possible failure of your work. For example, if a vase is not fired correctly, the glaze could craze and your vase will leak and, no longer being useable, will be discarded.

Worldwide, industry has always been aware of these problems, so it relies heavily on firing pots correctly to biscuit by using Bullers rings. These are direct decedents of Wedgwood's research into the art of pyrometry, proving that a controlled ceramic material shrinks at a set rate according to the amount of heat work. There are just three types of ring; low, medium and high, which can be placed throughout the kiln to measure heat variation from top to bottom. They can also be withdrawn during firing to measure progress and once the appropriate shrinkage is achieved the kiln can be turned off. Today this is an excellent way to test the accuracy of a kiln. In the past the gauge used to test the rings was prohibitively expensive, so it never really made its way into the studio pottery sector. Mantec, the company that supply Bullers rings, now provide a plastic version at a low cost.

CHAPTER 9

THE RIGHT STUFF

Potters are not always aware of the importance of the quality and suitability of their chosen clay body, the making process employed, and its suitability to the work they are producing. The tendency is often to go for the cheapest body available, but in most cases the potter's time is the most expensive component, so saving a few pence on clay can be a fool's errand.

I was once asked to be a consultant by a newly graduated potter who had recently set up his workshop. He was experiencing losses at all stages of making, with splitting, cracks and contamination often only showing after biscuit firing or even worse, after the glaze firing. It was a white firing body, so colour was very important. His loss factor was far too high, making it difficult to meet his orders and therefore his profit margin. As a guide, an acceptable loss from start to finish is about 10% and obviously very dependent on what you are making. As stated previously, there are always more losses with plates and dishes, but it was obvious to me that his problem lay with the clay body he was using. I explained that his only way forward was to change his clay supplier. I knew of a pretty much identical body available

Worcester coffee pot. Photo © Bouke de Vries

elsewhere but at a slightly higher price. He was very reluctant to change, not for a technical reason, but purely one of cost. I assured him that the working and fired quality were both superior to the body he was currently using. He had already looked at the body I was suggesting but was reluctant to pay the higher price and had decided to go for the cheaper alternative. I tried to explain that fewer losses would more than offset the cost and increase his production and profitability. It was not an easy sell; as can often happen when asking for advice, the answer is not always the one you want to hear. Often, it is the potter's lack of experience at the route of the problem. I experienced many occasions, when working at Harrison Mayer, where the problems were traced back to the potter, rather than the material. I only remember one instance where the material was at fault, when someone dispatched the wrong glaze.

But on this occasion the materials were at fault. The potter I was advising had a small amount of the old clay body in stock and assured me that he would make the change as soon as it was used up. Some weeks later he phoned me, informing me that he no longer had the problems, that his pots looked better than ever, and losses were next to nothing.

How does this fit into the green perspective? Follow the environmental trail. The wasted time, lighting, heating, firing, raw materials and disposal of biscuit and glazed pots to land fill, all an unnecessary waste of resources. In industrial ceramics, the sale of seconds can be the difference between making a profit and going out of business, but it is not always an option for some studio potters as they don't want faulty work reaching the market. I am always amazed by how much loss some potters will tolerate before seeking help. Even worse is an assumption that this is an acceptable part of a particular technique or process. If you are willing to improve your process, in most cases you do not have to compromise the aesthetics and qualities of your work.

I often reflect on my journey to becoming a potter. Four years at art school, four years working in studio potteries producing domestic ware and garden pots. Then moving to an industrial ceramic materials supplier for three years as a craft and technical advisor

'Atlas and the Broken World' by Bouke de Vries.. Photo © Bouke de Vries

to potters, schools and small industry, before setting up my first studio. I had the privilege of making most of my mistakes – and there were many – in someone else's time, and if it was not good enough it wasn't allowed through the system. The lessons came hard and fast in a studio that had to make a profit to survive. I spent some of my first week at one studio packing an electric biscuit kiln with pan-handled soup bowls and was very proud of myself when finished. I asked David, the studio owner, if it was done correctly. He smiled, then asked about the two left-over soups bowls on the ware board. When I said they wouldn't fit in, he asked me to unload the kiln and repack it until I could get them all in, which I said was a waste of my time. He replied that it was, but I would then learn how to pack a kiln properly as those two soup bowls would pay for the firing. It should be noted that no shelves or props were used in the biscuit kilns; all work was boxed and fired rim to rim, foot to foot. I sometimes see potters packing kilns where there is so much wasted space in both biscuit and glaze kilns.

When learning to throw, many students will refer to the 'individual' quality of their mugs, which I often find is code for not being able to throw two that are the same. Their opinion on this lack of uniformity soon changes when they have a kiln full of mugs that won't box, and half of the firing is wasted on kiln furniture and not their pottery. We will always find a good excuse for our own inability to do something well and cover up bad practise. Remember, a lot of energy is wasted firing the kiln furniture.

WASTE MATERIALS FROM CERAMIC PROCESSES

There is nothing new about the recycling of waste created by potters and the reintroduction of it back into the making process. Waste fired ware is an ideal aggregate as it is relatively inert. The fact that fired ceramics can withstand time, retaining colour and vibrancy even when broken, is relished by archaeologists. The most common form of recycled material is probably grog. This is a term for the addition of ground-fired ceramics to a clay body, a practice that has been accepted for as long as pottery has been produced. Grog is normally based on fire clay and added to stoneware and earthenware. Molochite is made from calcined china clay or occasionally ground HTI kiln bricks and is mainly used in white bodies. Sand can also be used as a cheap alternative, but too much may cause problems such as quartz inversion, leading to dunting. One or all of these can be added to clay to impart strength and workability, helping to cut down shrinkage and flabbiness, especially in larger pots.

There has been much research into the possibility of recycling materials to use as an addition to bodies and glaze.

We all like the idea of recycling in our day-to-day lives, but we need to look at how that can extend to ceramics. Up until you fire your pots, no matter how long they have been sitting on the shelf, you can reprocess them back to usable clay. A problem arises once it has been fired. In most cases, fired waste will go to landfill or is sometimes used in road construction, but the good thing is that fired waste is inert and does not normally cause contamination. In the past, broken pots, kiln furniture and saggars went to the shraff tip, some being crushed up for grog and added to fire clay to make more saggars and some used as aggregate in cement.

Skip full of contaminated white earthenware body destined for landfill; unfortunately it is more cost affective to discard it than reprocess it. In the past it could be sold or given away to other potters. Photo © Kevin Millward

Top left: scrap silicon carbide for crushing and recycling. Top right: bags of crushed refractories. Bottom left: skip of scrap white earthenware biscuit, thousands of tonnes discarded every year due to faults in the ware. Bottom right: Kiln props and shelves for crushing into grog.

Pins from plate cranks and assorted kiln furniture for recycling. Photos © Kevin Millward

What can a small studio or hobby potter do to help recycle their waste? Making your own aggregates is an option, if you are prepared for the work involved, crushing, milling and sieving your broken biscuit ware and adding it to your own clay body. Dedicated green potters have made it their mission to re-use as much as material as possible to prevent it going to landfill. Be aware that glazed ware does not make good grog unless any traces of sharp edges are removed. Plaster of Paris is a controlled substance and should never be disposed of with your household waste. It should be taken to your local recycling centre where you may be charged. Due to the large number of moulds used, factories arrange for a special skip collection, with the plaster usually being recycled to make plasterboard, preventing it going into landfill. An alternative use for worn-out plaster moulds is to break them up and mix with fresh plaster to make wedging and kneading bats. The quality and strength may be compromised, so it may not suit your needs. But be warned that little bits of plaster will cause lime popping, so do it outside, away from the workshop or studio. If you use waste plaster moulds in your mix, wet it first so the new mix sticks better when you pour it over the top.

CREATIVELY USING DAMAGED AND BROKEN POTS AND REFRACTORIES

In the past, it was a matter of course to repair or re-purpose pots that were very often expensive and considered purchases, rather than throw them away. Different cultures have found various ways of repairing and repurposing broken

High alumina kiln bricks for crushing and recycling, and used or damaged kiln props and shelves for crushing into grog.

ceramics. This was before super glue, two-part epoxy resins and modern fillers used by ceramic restorers. European ceramic restorers would use rivets for repairs, normally the preserve of itinerant craftsmen. Animal glue was used to hold the broken parts together and small holes drilled either side of the crack or broken section. Brass or copper rivets would then be put in the holes, much like modern staples; these were put in hot and would pull the broken parts together as they cooled. Some collectors like the qualities this type of honest repair brings to the pot.

The process of *kintsugi* is a Japanese technique used to repair and restore broken ceramics. It has become an art form and books have been written about the process and the philosophy behind it. As with most Japanese arts and crafts, an almost ritualistic approach to the preparation of the materials themselves has arisen, alongside the actual repairing of the pot, to the point where potters deliberately break their pots so they can repair them. There is a very good book on the subject by Bonnie Kemske called *The Poetic Mend*. Other artists and potters have embraced the broken pot and breathed new life into something that was, in most cases, considered fit for nothing but the dustbin. Bouke de Vries is a perfect example of an artist who has taken broken ceramics that would normally be discarded and re-assembled them into unique works of art, not just breathing new life into them but giving them a whole new existence; works of art in their own right, not intended to be functional. Bouke has no necessity to make or fire his work, and in many ways, he is the ultimate recycler of ceramics; not restoring them but reworking them into new creations.

Vast quantities of refractory waste are created on a weekly basis by industrial tableware factories, which the hobby or studio potter may not realise will more than likely end up as an ingredient in the clay body they use. A Stoke-on-Trent-based company has to hand select the different types of refractories, a task that should not be underestimated and that very few are willing or able to take on. The selected waste is crushed and wet milled into grog before being graded into specific sizes for use in a wide range of industries, and for most studio potters as an addition to a clay body. A brilliant example of re-using ceramic waste.

CHAPTER 10

FIFTY SHADES OF BROWN

Many potters try to attain a particular quality in their work that they feel can only be achieved through a specific way of firing. But that may tug at their conscience, due to its environmental impact. Can we condone burning large amounts of wood for the small number of pots produced, many of which may have to be refired, or can we use our knowledge to achieve the same results with a minimum effect on the environment? In a world where industrial processes have a dramatic effect on climate change, the long-term implications of burning fossil fuels – in particular wood – are making the practice untenable, so we urgently need to find a better way forward.

Within my own lifetime, I have witnessed the demise of coal-fired kilns following the passing of the Clean Air Act, and the introduction of natural gas from the North Sea, which replaced town gas produced by burning coal. In the first studio I worked at, the kilns were fired on oil using a small propane burner to establish a stable flame. In those days it was a cheap and convenient fuel and as the studio was in the countryside most

Experimental wood ash fired pot made by the author.
Photo © Kevin Millward

other fuels were not available. The fuel crisis in the 1970s put paid to low-cost oil, motivating most potters to change over to propane – ironically another petrochemical by-product.

Natural gas was not an option for most potters at this time as it was a relatively new fuel. Most domestic properties did not have an adequate gas supply to fuel a kiln of any size, so its use was usually restricted to industry and education.

Some consider that burning wood is sustainable and carbon neutral, as trees re-absorb carbon dioxide in the atmosphere. However, it must be remembered that it also releases many toxic and carcinogenic particles into the air. Ten tonnes of burned wood produces approximately ten kilograms of ash. Most of this ash remains in the fire box, very little lands on the pots. It will soon be unacceptable to cut down trees to burn without replacing them threefold, as old trees absorb much more carbon than new saplings.

As makers we have to be more mindful of our impact on the environment when adding salt, soda, propane, oil, gas and wood to fire. What is most important; the quality and integrity of the pot or merely the way it was fired? And burning tonnes of fuel in the hope that it will make a bad pot better is a hugely negative approach. We now need to think about how we can achieve the desired effects with the least detriment to our environment.

These questions set in motion a range of experimental firings to find out if it was possible to obtain the wood-fired effects with a minimum waste of fuel and loss of pots.

At Clay College we designed and built a series of kilns to investigate how firing with wood, to achieve a temperature high enough for the wood ash to flux with the body, can sometimes waste fuel. We wanted to see how much energy was wasted in the form of un-burned fuel and smoke and how much was used towards the heat work required for firing the pots. The main question we needed to answer was whether the act of firing with wood was being done simply for the experience, or to genuinely make good pots.

These effects can be achieved by less wasteful means, but do some potters feel that unless they are 'suffering for the

FIFTY SHADES OF BROWN

Experimental wood ash fired pot made by the author. Photo © Kevin Millward

Above and left: Experimental wood ash fired pots by Clay College student. Photos © Kevin Millward

cause', then it cannot be considered a proper firing? Unlike potters in the past, today we have a choice. Over the years, industry has adapted to rising fuel costs by developing more efficient insulating bricks, kiln furniture and forced air burners, using fuel more efficiently and cutting down on the length of firing. This was not always a concern for most craft potters, but it will be as the price of fuel rockets. In my experience potters would have previously built their kilns without considering which brick was the best for thermal efficiency. The use of underrated burners also meant a struggle to get to the desired temperatures, which led to the burning of extra fuel. A long firing doesn't necessarily mean a better firing. Our kiln at Clay College fires to cone 12 in only five hours with excellent results. I think the idea of having a long firing stems from the old, inefficient kilns, where pots were fired over six days in a bottle oven. However, they can now be fired in a modern kiln in less than 24 hours, achieving the same results without compromise.

BUILDING A SMALL WOOD KILN

Designing and building a small kiln that would fire using gas to provide the bulk of the heat work was quite a challenge. We side stoked the wood for reduction, introducing ash into the chamber, then blowing it into the kiln to simulate the taking up of ash from the fire box. We then fired up to cone 11, soaking out until the ash was fully melted. A small amount of salt can be added to mature the ash at your own discretion but is not necessary. Draw rings were taken out at different stages of the firing to ascertain the amount of ash landing on the pots and when it had fluxed.

At this point the students were split up into two groups, each one formulating a concept for a combination kiln design. The first group decided on a fire box for wood burning, backed up with a natural gas burner. We constructed it with HTI 26 bricks which gave it durability and a reasonable degree of thermal insulation. The fire box was loaded from the front and one gas burner was delivered to the side at the back of the fire box. There was no problem getting the kiln to temperature with gas but the wood, which had been donated or foraged locally, was poor and had little calorific value, contributing to little or no rise in temperature. The kiln produced little or no smoke during the wood-burning stages. We tried using a blower to encourage the fine ash from the ember bed to travel through the kiln chamber, but it was far too fierce and blasted in too much ash, so this was abandoned. Cone 10 was bending in under 12 hours, so the firing was concluded and due to the nature of construction and the type of bricks used, the kiln was cool enough to open the next day. The overall results were good and although some areas on the pots had more wood ash than others, they definitely had the qualities associated with wood firing and the loss rate was very low – possibly due to under firing or lack of ash. The conclusion was that using a fire box was not necessary, as the burners provided the heat work, and the small amount of ash pulled through could be delivered in a more effective way. The research undertaken during the development of this hybrid gas/wood ash kiln will have enabled the students of Clay College to explore the magical qualities of wood firing in an achievable and environmentally friendly way, having as little impact as possible.

The second group of students decided to build a kiln with a longer chamber that had two distinct areas, one for tumble stacking and the other for kiln shelves. A gas burner was placed at the back of the kiln and a small bag wall installed to prevent direct impingement of the flame. The ash was to be blown in from the back of the kiln over the burner port and flame with the objective of carrying the ash in a natural flame path through the kiln. The method of pushing the ash through the kiln was a complex one and we had learned from the previous build that the force of the blower was too great. We tried using a vacuum cleaner in reverse, a method I have successfully used many times before with forced-air burners; very effective but far too noisy. We finally decided to use cheap hair dryers, running on cold, attached to a stainless-steel tube connected to the kiln. We needed to find a way to put the ash into the hair dryer. The answer was to finely pass the ash through at least an 80s sieve to prevent blocking the hair dryer. Various methods of introducing the ash into the hair dryer were tried, and after putting the ash through the sieve, we found that using a small transparent plastic tub allowed us to observe the vortex created as the air and ash was pulled through the hair dryer. Other methods could have been considered, including a battery-powered alternative, but due to limited time the students opted to stick with what worked. The first firing of the second kiln proved extremely informative, with

Test wood ash kiln before and during test firing. Photos © Kevin Millward

Unloading test wood ash kiln Clay College. Photo © Kevin Millward

some fantastic results and some not so good. The main drawback was the uneven distribution of the ash. Unfluxed ash had built up on one side of the pots that were directly in front of the delivery port, but the bottom shelves had very little sign of ash. We decided to side stoke over the tumble stack area at the front of the kiln to create an ash build up on the pots, but this proved too much. The bundles of wood were made to fit the side stoke holes exactly and with careful adjustment of the flue no smoke was coming from them. I demonstrated this by using the passive damper to reduce the pull on the kiln chamber, releasing large amounts of smoke that disappeared immediately when it was replaced. As losses were so high, it was the consensus of the firing group that the tumble stacking was not a good idea and we discussed the best way forward with the knowledge they had gained so far. During the next firing we were side stoking, to aid reduction, as well as blowing in ash via multiple ports to get a more even distribution throughout the pack. Excellent conditions were created providing a good coverage of ash, even temperature and reduction, but the consensus was that we would no longer side stoke as it resulted in ash pooling in some of the bowls. We would rely solely on the gas and use the damper to induce the reduction. On our next firing we opened more ports on both sides and at the back of the kiln to create a more controlled ash deposit where students could place the pots in positions that would get the best results.

On the final firing a good reduction was achieved by using the damper with the addition of more gas. Once the kiln achieved 1220°C (2228°F), we blew ash in through the steel tube with an old hair dryer, using about 900 g (32 oz), evenly delivered through all the ports. Salt was delivered into the fire box when the kiln reached 1260°C (2300°F), in paper parcels of about 150 g (5 oz) – 750 g (26 oz) in total – then allowed to rise in temperature to 1280°C (2336°F) and soaked out until cone 11 was down and cone 12 was halfway down. Draw trials were also taken to see when the ash was fluxed and the firing was ended at this point, damper and burner ports closed, clamped up and left to cool. On opening the kiln the next day we were met with great results, the general opinion being that there may have been a touch too much salt on some of the pots at the front by the burners. Overall, it was a very even firing, in fact slightly hotter at the bottom as all cones were fully down, with good reduction and even distribution of ash.

In conclusion it appears that more than satisfactory results are possible, with very little loss, under what you may consider controllable and repeatable circumstances.

This way of firing could be a great opportunity for a potter just starting out, who has the ambition to wood fire but insufficient space and resources. They could create the quality and finish that wood firing offers in a low-cost, small-scale studio set up. Interestingly we have found when doing this type of kiln building with full time students and people attending workshops, that the surprise point was realising that it's not the build that takes the time, it's deciding what type of firing you want; whether you want to fire to earthenware temperatures, stoneware temperatures or do some reduction. Whether you want to use soda, salt, wood, or all three. All these options will involve choosing the appropriate type of bricks and burner system.

High temperature insulating brick HTI 34, cut to hold elements in top loading kiln. Off-cuts can be used to make molochite. Photo © Kevin Millward

Another surprise was how easy and gentle the firings were. The kiln was quick to reach temperature, gave even reduction, with no smoke and flames, other than when we pulled the bung to check the cones and a gentle flame from the flue and the small amount of gas used. Again, I demonstrated that putting in too much fuel is a waste, giving smoke and smell but not much else, other than slowing down the rate of clime. In most cases the glaze and firing results were excellent. I am still of the opinion that some potters are more interested in the spectacle, rather than making the process as efficient as possible and achieving some great pots as a result. Much research still needs to be done into improving firing processes and different methods of introducing the ash; it will be interesting to see what the future holds.

OXFORD UNIVERSITY KILNS AND CLAY COLLEGE

In the winter of 2023, the students at Clay College were invited to fire the newly-built kiln designed and constructed by Svend Bayer, operated by Oxford University. One of the sources of the wood on site had been planted in the past for future use in the furniture industry which, sadly, no longer exists. Much planting of trees was done in the early 20th century to future-proof British industries from the necessity of importing wood – particularly in times of war – but due to the long growth time, the requirements for this type of wood, or simply the demise of certain industries, has left vast swathes of woodland devoid of native trees. This wood now needs to be replaced by indigenous varieties, which means there is an ample supply of suitable wood for use in the kilns at Oxford, available on site with no transport costs.

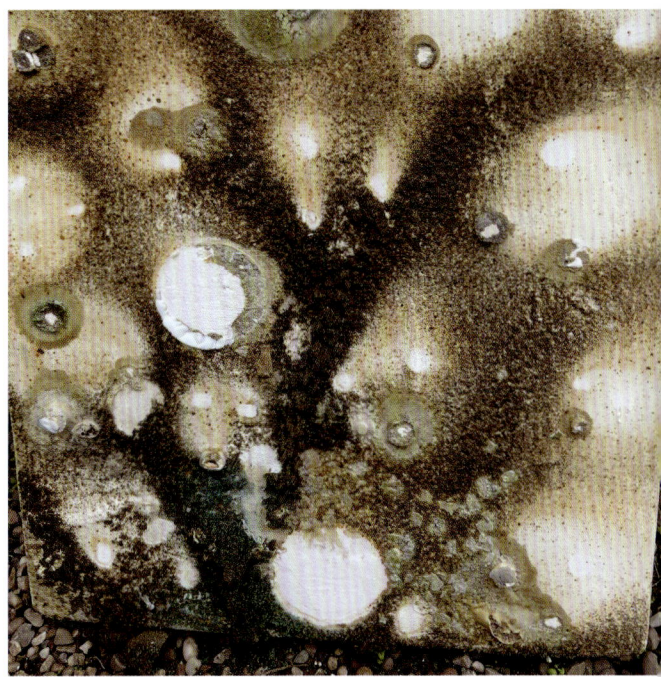

Wood ash on kiln shelf after firing. Photo © Kevin Millward

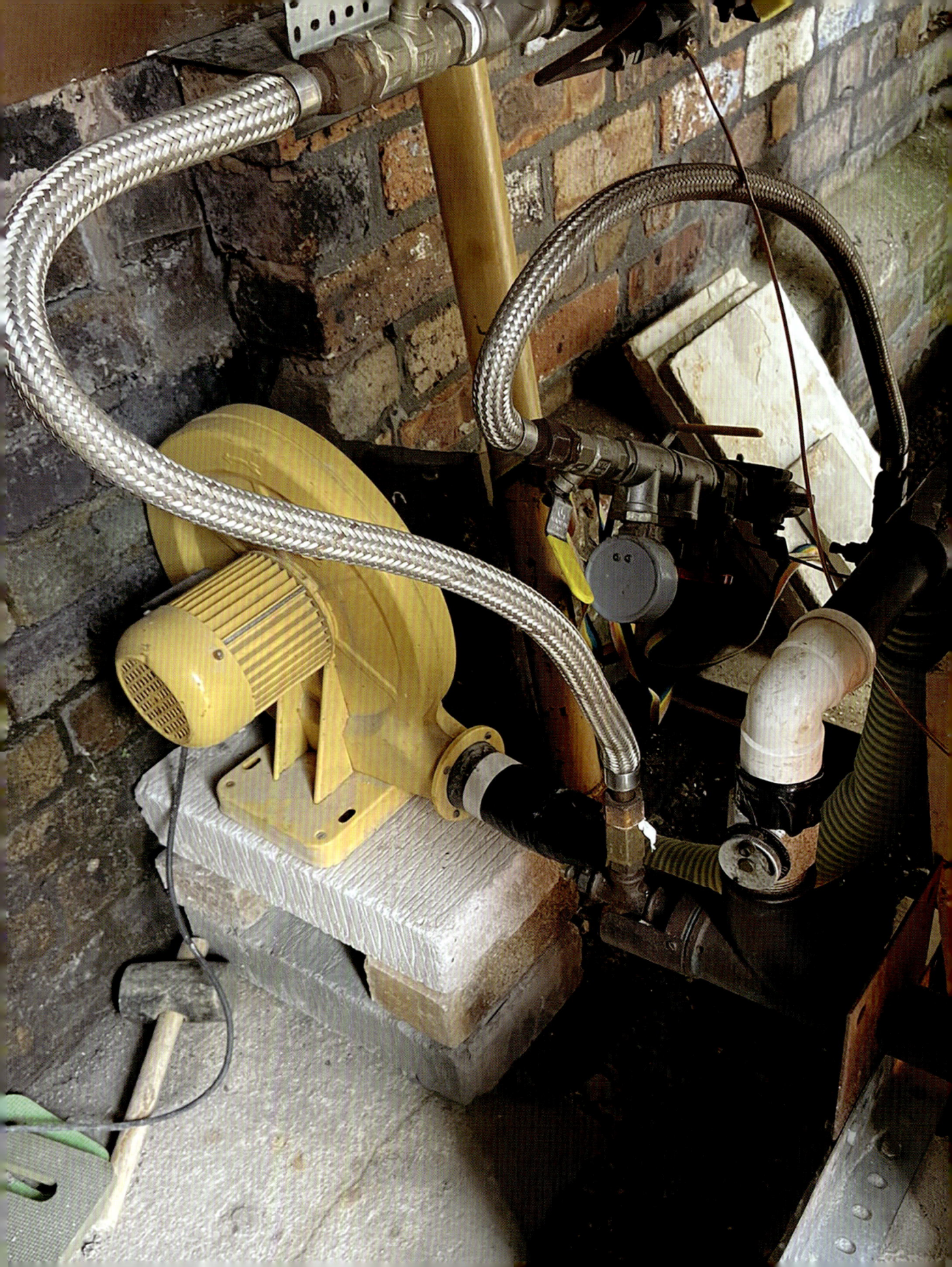

CHAPTER 11

BURNERS

As we know it is difficult to be completely green when we are firing pots, as it involves burning some type of fossil fuel. It is imperative that any fuels we use maximize every ounce of energy as efficiently as possible by using the most suitable type of burner system. Industry has a vested interest in finding the most efficient savings due to the high cost of fuel. Irrespective of which fuel you intend to use, your choice will be based on what temperature you are intending to fire to. The capacity of the kiln will determine the size and number of burners you will need. Whether you are wanting to once fire or biscuit fire will affect how many burners you will need to preheat the kiln.

We use three basic fuel types: propane, oil and natural gas, and the types of burners can be divided into two main types.

The first type is a naturally aspirated or Venturi burner using natural gas or oil that relies on the flue to create the draw of air into the burner. The faster the hot gases exit the flue, the more air is drawn in through the burner before mixing with the fuel. This can

Fan or blower for forced air burners. Photo © Kevin Millward

Above and opposite: Gas burners for natural gas, propane and small ones for raku. Photos © Kevin Millward

burner system off-putting. However, substituting it for a cheaper system necessitates a less economical, more polluting and longer firing. It is always better to overestimate the burner size, but if in doubt, provide your supplier with as much information as you can – kiln design, type of brick, firing temperatures – and they will calculate it for you.

Propane is a pressurized fuel and can be used with naturally aspirated burners that the kiln can run on with little or no flue, but the exhaust gases will have to be extracted, and the method will depend on the design and position of the kiln.

The least common in the studio sector are forced air burners. Using natural gas, propane or oil, this type of system is common in industry and can have fuel savings of as much as a third. The advantage of this system again is that a flue is not needed to provide draw, only to take away the waste gases, so does not have to be vertical or attached to the kiln. The air and fuel are mixed to obtain the most efficient combustion, but if you want to create a reduction atmosphere the amount of air mixing with the fuel has to be reduced, whilst the heat in the kiln

often involve using a tall, vertical flue to provide enough draw of air to burn the fuel efficiently and the design of both the kiln and burner system will determine just how efficiently this happens. A common error is to underestimate how many Btus (British thermal units) are required, which can make the initial cost of the

is high enough to maintain combustion. Personally, I balance the amount of air-to-fuel mix by also restricting the damper on the flue to create a perfect reduction atmosphere, while maintaining a steady rise in temperature. I have used this type of burner system all my potting life, first with oil and then with natural gas. These kilns are fast, efficient and fire very evenly, having the ability to crash cool if required. Forced air burners are more expensive initially and require an electricity supply, but these costs are soon recouped by the amount of fuel saved. Kiln building companies use what is referred to as a package burner, an integrated safety system providing spark ignition, flame failure and a timed purge that prevents a build-up of gas in the kiln chamber. Older systems require the door to be left open until the burners are established, after which the door can be closed. Package burners can also be connected to a program controller that operates valves to increase or decrease the gas input. Reduction is usually controlled manually.

One type of burner I have not mentioned is a drip plate which is used mainly with waste oil. Fuel oil is simply dripped onto a preheated metal plate using a small gas burner until the metal plate is self-sustaining and as it doesn't have to go through a fine hole, the oil does not need to be finely filtered.

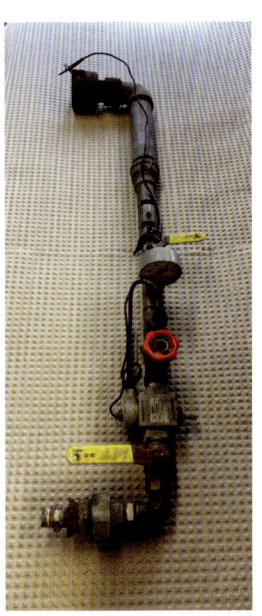
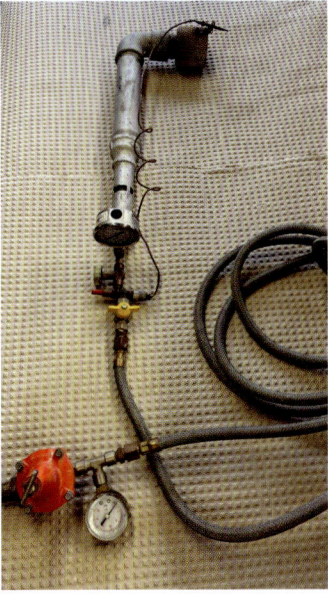

CHAPTER 12

GLAZE APPLICATION

Glazes can be broken down into four basic types: earthenware based on lead; borax and alkaline frits; stoneware glazes based on feldspars and wood ash; and vapour glazes based on soda and salt.

After more than 50 years of making pots for myself, teaching at numerous colleges and universities, running workshops and seminars, and being a technical advisor for one of the UK's main ceramics suppliers, I can say with confidence that more pots are ruined at the glazing stage than at any other part of the production process. A huge amount of over-glazed, under-glazed, over-fired, or under-fired work from the glazing room ends up in the skip.

I was part of a team in the 1970s that investigated how we could encourage more people to take up pottery as a hobby. The three things that were top of the list that prevented people taking the hobby out of the classroom were needing to have access to a potter's wheel, a kiln at a price that was affordable and could be used at home, and the skills for glazing pots without ruining them. This period saw the

Dipping a biscuit-fired pot. Photo © Kevin Millward

introduction of brush-on glazes, which meant not having to buy vast amounts of glaze, and an almost foolproof way of getting the correct amount of glaze on your pot. Great for the amateur home potter, but not always the solution for the aspiring professional. I hope this chapter will help clarify some of the dark art of glaze application.

The physical application of a glaze to a pot can be traumatic for most potters, with the realization that if you make a mistake with the glazing it can mean disaster. It is difficult enough to think about the aesthetic considerations of the glaze and what it will hopefully look like, let alone the physical application of the glaze. Most experienced potters have honed the craft of applying the glaze by trial and error. This however does not help the potter who is just starting out. How do you set about working out the best method for the application of the glaze in your situation? You will need some rules to work with. As the setting up of a glaze before application is so rarely taught today, some questions to consider are how much glaze should be put on a pot, how thick the glaze should be and how that might vary according to the glaze and the type of body it is

applied to. The correct thickness can only be established after testing things such as the biscuit temperature the pot was fire to, and the pint weight the glaze was set to. I would suggest if you are unsure about the glaze thickness, fire pot on a piece of scrap kiln shelf.

If you are just starting to glaze on your own for the first time and have no one at hand to help, this is how you would commence making your glaze.

Always put the water in the container first, then add the dry glaze materials to the water, so as not to create dust. This can be done under extraction or by using a damp cloth over the container. It is better to use too much water as this will aid the mixing of the glaze. When fully mixed you should sieve the glaze. I would recommend you use at least a 100s mesh and pass the glaze through twice. The first time you can use a brush to help the lumps of glaze through, but do not force the glaze on the second pass, as this can create lumps. Now let the glaze settle to remove any excess water. You can now set the pint weight of the glaze; this will determine the amount of glaze to be deposited on the surface of the item. The pint weight refers to the

GLAZE APPLICATION

amount of dry glaze suspended in a pint of water. This information can be used in conjunction with Brongniart's formula to calculate the dry addition of oxides or stains to a slop glaze. The weight of a glaze is specific to the size and shape of the pot, and most importantly what biscuit temperature it was fired to and its porosity or lack of it. As a rough guide, the higher the biscuit temperature, the higher the pint weight. For example, a standard stoneware glaze that you might dip onto a mug that has been biscuit fired at 1000°C (1832°F) would be approximately 907 g (32 oz) for one UK pint of glaze. If you dip this glaze onto the pot, glaze fire it and the glaze is too thin, you will know that you need to increase the weight to say 935 g (33 oz) per UK pint. This would involve the removal of water or the addition of more glaze material. If the glaze is too thick, then you would need to reduce the weight to say 879 g (31 oz) which would be achieved by the addition of more water. To set up to use a pint weight, simply take a pint container such as an old glass milk bottle or a pint can from your supplier. First counterbalance the container on your scales so that it weighs zero. Now stir the glaze to make sure the ingredients are well mixed. Decant one pint of glaze into your milk bottle or container and place on the scales. This will give the weight of the glaze in grams or ounces. With this information you can adjust the glaze. Without this technique you would not know how much glaze you had suspended in the water. The pint weight can be used for all methods of glaze application. If you have problems with the settling out of a glaze or with glaze running, a flocculant can be added to the glaze. This will prevent settling and improve the way a glaze holds on the surface of a pot. With the right pint weight and the addition of enough flocculants you can dip biscuit ware that has been vitrified, such as bone china that is biscuit fired at 1280°C (2336°F). The two most commonly used flocculants are Epsom salts and calcium chloride. Many potters have used the addition of a clay such as bentonite to do a similar job, but the disadvantage with this is you cannot remove it, as it becomes part of the glaze mix, whereas flocculants can be removed as they are soluble in water. By adding water to the glaze and letting the glaze settle, you can then decant any excess water off, and you will remove a quantity of flocculants along with it, then replace with fresh water. It is possible to measure the fluidity of the glaze using a torsion viscometer, but as most potters don't have access to one, an alternative is to use a hydrometer. Many potters use them instead of the pint weight system. I would personally recommend first using the correct pint weight and then setting up the flocculant with the hydrometer. The use of one or more of these systems can prevent many of the problems associated with glazing.

Another common problem when glazing is the propensity for glazes to be knocked off along the edge, so the addition of a glaze binder such as gum Arabic or a preparatory glaze binder from your supplier can help to alleviate this problem.

BRUSH-ON GLAZES

The brush-on glaze is a more recent addition to the techniques of glaze application. Brush-on glazes solve many problems associated with glaze application, especially for the beginner, as they are available in small quantities and the range of colours, effects, and firing temperatures is extensive. It is possible to brush the glaze on to the biscuit ware without it instantly drying on the surface because the glaze has Carboxymethyl cellulose (CMC) added to it, which prevents the glaze from drying too fast by holding onto the water and allowing the surface tension of the liquid to level the glaze and remove brush marks. The only downside is that you need to build up multiple layers of glaze as the addition of the CMC bulks out the glaze too much for it to be put down in one layer. If you don't want to buy ready-made glazes you can add CMC to your own glazes. Contact your suppler for more information. It can also make a very useful addition to small amounts of glaze used for repairing a glaze surface that has faults before re-firing. Some potters are still resistant to buying glazes but remember, it's what you do with it that's important.

SPRAYING

Spraying has some great advantages but unfortunately has a major drawback in that it is only realistically possible if you have a spray booth, compressor and a spray gun. I have seen some potters use crude and slightly Heath Robinson-esque apparatus, such as cardboard boxes and vacuum cleaners for spraying into.

GLAZE APPLICATION

Spray booths for most potters are a luxury they can't afford, but you can build your own. In the UK, health and safety regulations specify that the air flow into the booth must be at a specific rate to prevent fine glaze dust being inhaled by the potter. A constant flow of water catches most of the over spray in the water. This is referred to as a wet back spray booth. If you are just using the same glaze on a regular basis, you can reclaim the over-sprayed glaze. Any glaze that is not caught by the water is captured by a bag on the back of the spray booth and prevents any dust entering the atmosphere. You should not vent any glaze dust directly into the atmosphere. It is unacceptable to put in place all the measures to prevent the potter inhaling glaze dust, just to dump it on someone else.

I remember going to visit a young potter who had set up their studio in an arts centre in a major UK city. They mainly sprayed their glaze, as they were not confident with dipping their pots. Many students spray their pots at college as this means not having to mix large amounts of glaze. The potter showed me with great pride the spray booth they had constructed themselves. It worked

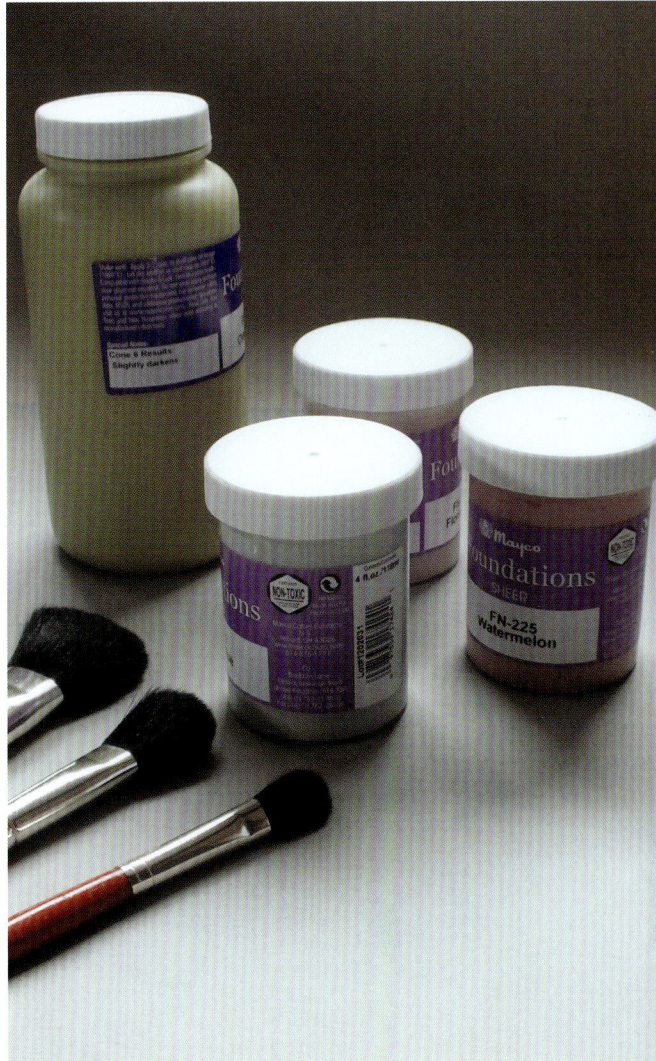

Brushes and pots of brush onglaze and colour.

very well for them, pulling the over spray away from the potter. However, when I asked what sort of filter they had and where it was being vented to, it transpired that there was no filter as they felt it restricted the air flow; as for venting, they were using a large flexible tube fed out through a window, much like the ones used on a tumbler dryer.

On leaving the studio, I decided to explore the side street adjacent to the studio and discovered that there was a large stripe of glaze attached to the wall downwind of the tube. Either the potter was totally unaware or was unbothered that the very dangerous fine silica dust was being blown directly into a street used by pedestrians. It is very often the case that outside an area where pots are produced industrially, potters are unaware or unbothered about health and safety regulations and some inspectors are also unaware. For example, one rule under the health and safety regulations for pottery studios states that nothing should be placed directly on the floor that would prevent or affect wet cleaning the entire floor.

Silica dust is a major concern in industry and all possible efforts are made to prevent its transmission. Workers are instructed to change into protective clothing for working in, including footwear. They have lockers for their workwear and lockers for their outside clothing, so as not to transfer dust from work to their homes.

Part of the problem with asbestos was that workers would transfer the asbestos dust to their homes, in some cases affecting their families. Whilst working in ceramics is nowhere near as dangerous as working with asbestos, silica dust is certainly not to be underestimated as

Wet back spray booth does not expel over spray to atmosphere but reclaims if required. Photo © Gladstone Enginering

a potential hazard. Be aware of the dangers involved, as the only safe and effective solution today is a wet back spray booth. It is not advisable to expel glaze dust into the atmosphere, as you do not know who or what the glaze dust may land on. If you are going to use the spraying method as your main method of glaze application, I would strongly advise you to also use an approved dust mask. This which may seem excessive, but it is better to be safe than sorry.

With the correct type of spray booth in place, you will also need a compressor and a spray gun, which used to be very expensive but are no longer excessively so. I would suggest not buying a very expensive spray gun as the cheap ones do the same job. The internal parts of most guns wear out quickly due to the abrasive nature of the glaze, so it is often cheaper to buy a new gun than attempting to repair it. They are mainly steel and aluminium, so can easily be recycled.

You will need to set up the glaze with a pint weight of about 964 g (34 oz) and I would also recommend adding a vegetable dye to the glaze so that you can see where you have sprayed.

This can really help, especially if you are spraying white glaze on a white pot. Add the flocculants to prevent the glaze settling out in the gun and prevent glaze runs on the pot. The glaze mix can now be passed through a minimum of a 100s mesh to remove anything that may block the gun and prevent it delivering the volume of glaze required. If you are using glazes that are made from wood ash this can present a problem for some glaze guns. I have found old worn-out guns are good for this type of glaze. I would recommend passing the glaze through the sieve twice; the first time you can use a brush to help the glaze to pass through and break up any lumps. Pass it through a second time without assistance. Spraying glazes with a built-in speckle can present a problem, as they may be sieved out or they could block the gun. Again, a worn-out spray gun and not sieving would be good to use for this type of glaze. It is normal to pour the inside of the pot, as attempting to spray the inside of some shapes can cause the glaze to blow back in your face. I recommend spraying the base first (if it is to be glazed), then place on a stilt so the pot is elevated to prevent build-up of glaze around the base, and then spray around where handles etc. join the pot.

Now the main body of the pot can be sprayed making sure the gun is set up correctly. Most guns have the facility to change the spray pattern, so consult the manufacturer's instructions. If you spray too close, you can over-wet the surface and blow glaze off the pot, but too far away and the glaze can have a dusty appearance. It may be necessary to apply multiple coats to avoid getting the glaze surface too wet and creating runs. If you find you need to re-spray a pot that has been gloss fired, you will need to set a higher pint weight and add more of the flocculants. Warming the pot is not always necessary providing the glaze is set up correctly and it is also a waste of energy.

For many potters, the area that gives them most cause for concern is how thick the glaze is. Cutting the glaze with a fingernail will give an indication as to glaze thickness, and you should always be aware of over glazing the pot as this can cause problems. There is in fact a tool for testing the thickness of a glaze application called a penetrometer. It consists of four small metal wheels; the two outside wheels run on the surface of the pot cutting through the glaze. The other two wheels glance the glaze surface, one at ten thousandths of an inch the other at twelve thousandths of an inch. It is always easier to put more on but it's a bit more complicated to get too much glaze off!

The advantage of spraying is that you can glaze a wide range of both simple and complex forms without having to make up vast quantities of glaze, and with good housekeeping, you can reclaim most of the over spray when using a wet back spray booth. This equipment can also be used to spray a whole range of ceramic materials such as onglaze colours, under-glaze colours and lustres, using a smaller and finer spray gun, usually referred to as aerographing.

TIP: If you are spraying dark coloured glazes, it is best to have two guns; one for coloured glazes and slips and one for white glazes and slips, to prevent cross contamination, especially in the case of cobalt!

DIPPING

Dipping is probably the most efficient way to obtain the high-quality finish required and is the most commonly used method of glaze application.

Unfortunately, for many potters today, it has one major drawback; the amount of glaze required to dip the pot into can be many litres, so the bigger the pot, the bigger the buckets and the more glaze you require. Most of this glaze will never be used. So the first consideration is having enough glaze and a container big enough to take any displacement should there be any. Items such as large narrow-necked vases, where the inside has been poured and the outside needs to be dipped, can displace a large amount of glaze.

Once you have your glaze mixed up and you have the quantity required, you can now set the glaze up to the correct pint weight. This you will have previously determined by testing, e.g., a high pint weight for the smallest pots that you will dip first. When the smallest pots are done, you can reduce the pint weight for the next size of pot. This is done for each size of pot up to the lowest pint weight for the largest item.

You will find, as you gain experience in glazing, that you will get a feel for the glaze and the consistency that you are using, so much so that the reliance on the use of the pint weight will wane.

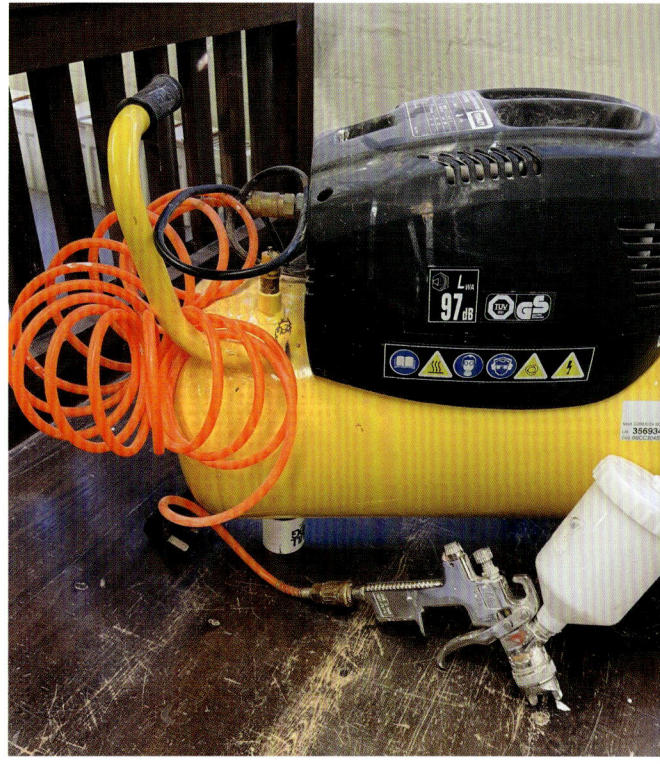

Spray gun and compressor. Photo © Kevin Millward

But you should never become complacent, as it can come back and bite you.

The way the pot is held during the dipping process can give great cause for concern, especially where the fingers physically come into contact with the pot. There are a few ways of getting over this problem, by using dipping tongs, claws or finger stalls. The next thing to work out is how to get the pot into the glaze, how you will get it out, and whether there is enough space to rotate the pot in the container to allow the glaze inside to pour out. Otherwise, you may have to lift the pot full of glaze out of the container in order to empty it, which is not easy if it is a big pot. Don't forget that all the time the pot is in the glaze, the amount of glaze taken up is increasing (another reason for the use of high biscuit in industry).

Once you remove the pot from the glaze you need to consider where to put the pot down. You would be surprised how many people find they have dipped the pot before thinking about where to place it afterwards. It is a good idea to rehearse the process before committing the pot to the glaze.

POURING

The advantage of this technique is that is does not require large amounts of glaze, as with dipping, but does require a degree of skill in preparing the equipment and in setting up the glaze. Sieving and setting the correct pint weight are essential. I usually go for about 907 g (32 oz), but if it's a very big pot I may go as low as 822 g (29 oz). This could be a better option as you will be less likely to apply too much glaze. In most cases, the glaze is applied to the inside of the pot first. Be sure to clean off any overspill first and be aware of not over wetting the pot when removing the glaze, as this will compromise the porosity of the rim, especially if it has a very thin edge. If in doubt, leave it to dry off. Now you will have to support the pot in order to pour the glaze over the outside. It may be necessary to wax the base, or if there is a deep foot ring you will have to blow out the glaze from the recess. This must be done while the glaze is still wet, so as not to form runs or drips. If you wish the foot to remain glazed, you can leave it to dry. Once dry, remove the glaze using a wet sponge.

I recommend using a bowl, bucket or some type of receptacle to receive the glaze, placed on the whirler. I have been known to use a plastic dustbin lid for large dishes and bowls. You can use sticks of some kind on which to rest the neck or rim of the pot (sticks with a thin edge are better), taking care not to chip the rim. Remember to be careful, if the pot has a narrow top or neck, that it does not topple over. You can place a stick inside the neck if you think the pot is unstable. For this you can use a lump of stiff plastic clay to put the stick in, then place this in the receptacle you are using to catch the glaze. You can retrieve the clay after washing off the glaze. Make sure the glaze is dry to prevent damaging it, before removing the pot from the apparatus. If you find you have too much glaze on the top edge, you may have to rub this back to an acceptable level. Remember to wear a mask or work under extraction because of the dust.

VAPOUR GLAZING

This process is not commonly used, due to the type of kiln required and the appropriate place to build and fire it. There are many ways to introduce the salt or soda, and potters are always coming up with ingenious methods to do it. But to keep it simple, salt can go in dry where it will volatilize and combine with the silica in the body to form a fairly even glaze throughout the kiln. Soda is normal combined with water enabling the soda to break down and bond with the silica to form the glaze. This method is considered to be more directional. Some potters make up paper parcels to introduce the salt, wetting them before putting them into the kiln. Be aware some paper has a high china clay content so can contaminate your pots with white specks, so find a low clay paper. Remember these glazes are clear and any colour comes from the body or slips applied to the surface of the pot. The insides of the pots are glazed with a liner glaze, as the salt and soda don't tend to get inside the pots. These processes can be used very successfully with wood firing and / or wood ash glazes. But you may want to consider the environmental impact of this way of working, it's not for everyone.

TIP: You can take all your left-over glaze dregs and mix them together. Obviously keep stoneware with stoneware and earthenware with earthenware. We did this at art school and for some reason the stoneware one nearly always fired to a blue colour. If you have any flocculants added, such as Epsom salts, calcium chloride or even CMC, these can be flushed out by adding lots of water as they are soluble. The water can then be decanted off, so you could also reclaim brush-on glazes in the same mix. Be warned, you must test them just in case they are very fluxie.

CHAPTER 13

PACKING AND DISPATCHING

Packing and dispatching the finished product to your customer cannot be avoided unless you only sell directly, and even then, it must be wrapped to protect it. Personally, I am not an advocate of old newspaper, second-hand bubble wrap and old plastic bags. When a customer has spent an often-considerable amount of money on a special purchase, the least we can do is wrap it properly and bag it nicely.

Japanese potters take great pride in the packing and presentation of their pots before sale. I am not necessarily in favour of silk bags and wooden boxes, but good presentation is important. So how do we do this without an impact on the planet? I personally use tissue paper, bubble wrap made from recycled plastic bottles and paper carrier bags, and I include a business card or postcard (cardboard not plastic) with my contact information. I hope that if bought as a present this packaging will be re-used. If you don't want to use bubble wrap you could use crinkled

Recycled cardboard box turned inside out to present a more professional appearance. Photo © Kevin Millward

brown paper. It is really down to your own conscience how you wish to present your work.

When packing for dispatch by a courier, this means handing over responsibility for your work to a third party, who may not have the same level of respect for it as you, the maker, so it is important to pack the work properly to give it the best possible chance of arriving intact. Most carriers will not insure the work against breakage and even if they do it usually only covers the making cost. They will insist that you use new packing materials, corrugated, double-walled boxes marked 'fragile', and pack items correctly. If this is not done, they may refuse to pay compensation if there is damage.

We need to consider the impact of plastic-based packing materials on the environment compared to those based on paper and cardboard and how effectively they can be recycled, if at all. Looking back at my first job after leaving art school in 1973, which was in a small studio pottery producing domestic stoneware, the least popular task was packing. We used the most recyclable materials of the day; old tea chests, newspaper and straw, possibly as green as you can get. Breakages in transit was virtually unheard of. However, the drawback was that newsprint blackened our hands, the straw was scratchy and the sharp metal strips on the edges of the tea chests cut our hands. Not very user friendly! Another popular packing material then was wood wool (shredded wood), used in the same way as we use polystyrene chips today. Many of the old types of packing materials were green, but were heavy, so more costly to transport.

To make the least environmental impact with your packing and delivery the lightest, greenest packing materials need to be used, together with transportation by rail and electric-powered delivery vans. Remember that breakage in transit means all the energy used to produce the item, pack and deliver it is wasted, so it is essential this is done correctly to ensure the goods are delivered undamaged. I am often bemused by the way some potters pack their work. I have, for example, seen pots worth thousands of pounds, having taken months of work, then wrapped in old clothes, newspaper and string, packed in vegetable boxes from the local supermarket, all tied together with old rope. It was a miracle the pots arrived in one piece!

The ceramics and transport industries have together developed a system for packing and delivering that no longer involves large amounts of packaging materials. Instead, they have set sizes of interlocking cardboard trays for the ware, in differing depths, with a fixed delivery cost irrespective of their weight. These also fit perfectly on a standard European pallet, a system I have personally used when making garden pots for the John Lewis Partnership, with never a single breakage.

Paper and cardboard are not always as green as you might imagine. The amount of energy, clean water and toxic chemicals used to break down the pulp together with bleach, to whiten it to the expected quality, is far greater than that used to produce plastic.

Sadly, irresponsible disposal of plastic releases it into the environment and therefore creates a huge problem, which includes the exportation of UK waste to the third world, where it is very often dumped into landfill sites and finds its way into the water course and oceans. Plastic itself has many important uses in our lives, but the problem lies in how we humans fail to dispose of and recycle it correctly. There are many opportunities to re-use packing materials like bubble wrap, polystyrene chips or the organic and biodegradable versions. Be warned though; I started to use starch-based infill, as opposed to polystyrene, but the local mice and rats loved it, so I stopped using it. Much packaging is commonly thrown away by shops and businesses. If sending work out for sale or return, the shop or gallery can be asked to save the packing materials to return any unsold items or use it themselves. Good cardboard boxes can be used again if in good condition. By turning them inside out, any logos or markings are inside, and they are as good as new. If they are too big, they can be re-sized with some cutting and folding with a craft knife, giving an old, oversized box a new life. Offcuts of cardboard can be used inside as added protection.

TIP: Work out the weight difference and costings of your packing materials and their environmental footprint against the added shipping cost. Use good, clean, used packing materials if you can. In the past, when straw was used as a packing material for export, it required a veterinary certificate to prove it was clean and free from pests and disease.

CHAPTER 14

HOW CLEAN IS OUR FUEL?

What are our fuel options today? Even if we all only used electricity, the vast percentage of this is derived from coal and gas. Below is a breakdown of the main fuels that potters have available to them, with their emissions of carbon dioxide (CO_2) into the atmosphere.

WOOD

1 kg (2.2 lb) of wood contains 450 g (15.9 oz) to 500 g (1.6 lb) of carbon, meaning that 1 kg (2.2 lb) of wood is holding about 1.65 kg (3.6 lb) to 1.80 kg (3.9 lb) of CO_2, which will be released on burning. Wood is not regarded as a fossil fuel as it is expected that new growth of wood will absorb this carbon. However, burning it still releases high levels of particulates into the atmosphere that can be extremely dangerous to health, especially to the lungs.

PROPANE

As it is not practical to weigh this type of fuel, we will work in Btu per million to compare the different

Wood-fired kiln under construction using fire bricks and HTI 38 bricks Clay College. Photo © Kevin Millward

fossil fuels. For example, burning one million Btu of propane will release 63 kg (139 lb) of CO_2 into the atmosphere. Propane is considered a clean fuel as it does not put many particulates into the atmosphere.

DOMESTIC FUEL OIL

This used to be the fuel of choice for many potters and was easy to store, but it became unpopular due to rising oil prices. One million Btu of fuel oil, when burned, will produce 73 kg (161 lb) of CO_2. One area of concern with this fuel, as well it being a fossil fuel, are the particulates it puts into the atmosphere.

NATURAL GAS

This is considered as one of the cleaner fossil fuels, it contributes virtually no particulates into the atmosphere. It has the added bonus of an integrated distribution system of pipes that delivers it directly to your home or studio. There are no road miles, as with propane and fuel oil, and you can 'pay as you go' as opposed to in advance. One million Btu of fuel will release 53 kg (117.0 lb) of CO_2 into the atmosphere.

HYDROGEN

As a viable fuel for potters, hydrogen is a long way off yet, but when it does become available it will be a game changer. It's a clean fuel as when it is burnt it turns back to water. However, it is currently made by burning fossil fuel.

BROWN'S GAS

Brown's gas is a hydrogen/oxygen mix produced by electrolysis. The process involves splitting water into its component parts, which on combustion, then returns to water. You can build this system yourself or you can buy a Brown's gas generator. I have not calculated how much energy, relative to the amount of gas produced, would be enough to fire a decent sized kiln. This, however, could be a way forward, using electricity to enable gas firing and reduction with a low environmental impact. These machines are commonly used in the third world for welding.

ELECTRICITY

Wind and solar are, in essence, very clean fuels with virtually no carbon footprint resulting from their production.

However, they create a vast carbon footprint due to the fossil fuels used during the construction of their infrastructure and there are many problems with non-recyclable waste such as the glass fibre and resins used in the blades of wind turbines, that have a limited life of about twenty years. Currently these are not recyclable. There are, however, moves to grind them up and burn them as fuel for making cement.

It will be a long time before we can produce enough electricity by wind, water, and solar power to be fully carbon neutral. Electricity created by burning fossil fuels is here for some time, but remember, it can be burned more efficiently in a power station than in a domestic setting. Nuclear power will still be a major player but replaces CO_2 with radioactive waste; that is of course until they crack fusion, which is, according to the experts, not far off.

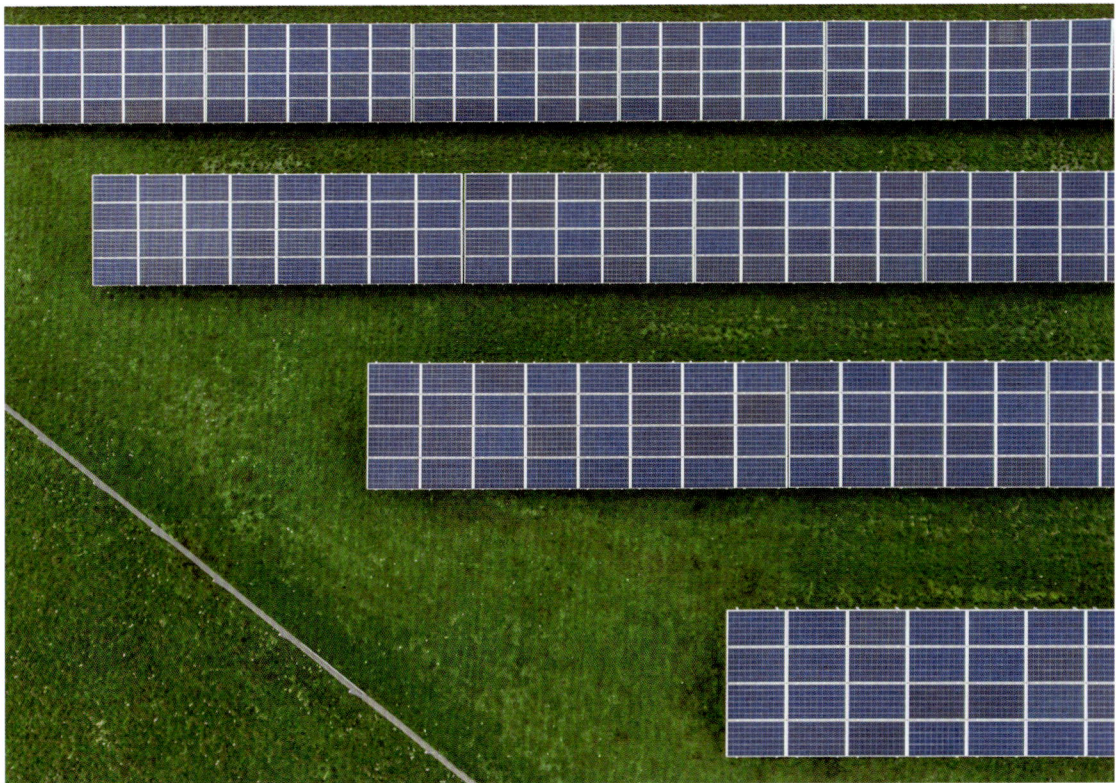

Solar panels generating electricity can be used to fire small kilns or stored in batteries for later use.
Photo © Getty images

SOLAR STORAGE

Sarah Mook is based in Oxford. Whilst having her green eco-house built, it was an ideal chance to incorporate sufficient solar panels and a Tesla power wall, to provide enough power to run the home and a small kiln. She has a 13.5 kw Tesla battery fed by 14 kw of solar panels, orientated to east and west due to planning restrictions. The kiln itself pulls about 7 kw of power which the battery and direct feed from the solar panels can provide. If the amount of sun is less and the battery capacity becomes insufficient to complete the firing, it can draw power from the grid to supplement this and finish off the firing. It must be acknowledged that this system was part of the building plans for a new home and therefore the ability to fire a kiln was a bonus. To retrospectively fit this system may not be financially viable in the short term.

FUEL FROM SEWAGE

This is something new on the market and was developed to help replace jet fuel, which is a type of kerosene similar to paraffin. This could be used as a replacement for fuel oil and is carbon

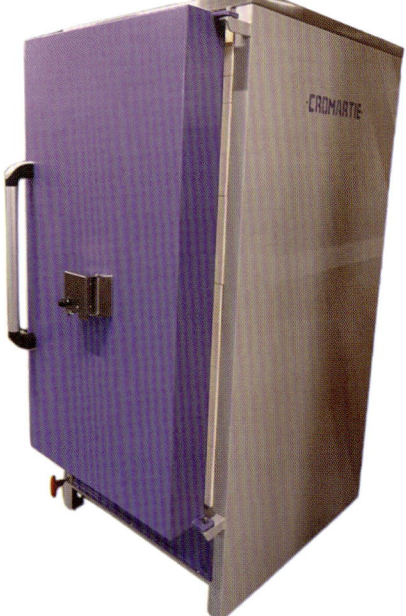

Modern front loading kilns. Photo © Cromartie Kilns

free. The cost is yet to be confined, but it could be another game changer.

ELECTRIC KILNS OLD AND NEW

The pioneer potters would have built their own kilns firing on wood, oil or gas if it was available. Some studios had old electric kilns, often second hand from small industry and usually requiring a three-phase power supply, not normally available to the home potter. There was a movement away from the constraints of the pottery evening class and a desire to set up pottery studios at home. Small kilns were available, but these were just a scaled-down version of their industrial cousins. What changed the face of pottery in the UK was the introduction of the lightweight, top-loading kilns from the USA, powered by a domestic electricity supply. It didn't take long for the UK-based kiln builders to catch on and similar kilns built in the UK were soon available. These kilns can be up to 0.25 m^3 (9 ft^3) and are connected to a 40-amp outlet, the same as a cooker. They fire fast, cool fast and now come with simple-to-use controllers, some being able to connect to your smartphone so you can track your firing remotely. I am not saying that the traditional front loader is dead; it has its place, but without the option of the top loader, many potters would not have got started. New or second-hand front loaders are an option for many but can be more costly to fire and take longer to cool down. For anyone thinking of buying an old kiln and rebuilding or repairing it, this can be a great way to get a low-cost kiln, but be aware, as some very old ones may contain asbestos which requires specialised disposal, making this option uneconomical and hazardous for some potters. Cutting bricks for electric kilns can be complicated as the element groove can be tricky if you don't have the correct equipment. I once re-lined an electric kiln but after much consideration decided it was wiser to get the bricks cut for me and together with a new set of elements, the result, in effect, was a new kiln.

TIP: It is possible to wind your own elements, but it's not practical for most potters, especially for those using top loaders as the elements are so long. You would require a mandrel or wooden dowel the same diameter of the element, and some form of lathe or electric drill to rotate the dowel. You would need to know the gauge of the

Kanthal wire and how many winds there are on that bank of elements. It is not something most potters want to take on, but it is possible; I have done it for a small front-loading kiln at a fraction of the price of pre-wound ones. Warning: there can be a recoil of the wire when wound, so be careful.

With advances being made in the efficiency of small electric kilns, potters are turning to solar power to either offset the cost of firing or totally cover it. The possibility of storing electricity, say with a Tesla power wall, directly from the solar panels when the weather allows, or feeding excess power into the grid, could offset the cost of firing. One concern is the move to reduce the standard power supply available to new-build homes. Currently most domestic supplies to older properties are 100 amps which normally gives the home potter 40 amps to use for a kiln, leaving 60 amps to supply the home. It has been proposed to cut this down to a basic supply of 60 amps, which is fine for most domestic uses but would make it difficult to run anything but a very small kiln. The good news is that most rented studio spaces designed for makers usually have a three-phase supply as standard.

There is an assumption that once firing can save money because you are not biscuit firing the ware, but this is not always the case. Traditionally all pots were once fired. Glaze, if any, was applied to the clay ware. When dried, the ware was then fired, maturing the glaze and body together. With the development of more formulated white bodies such as white and cream earthenware and the introduction of bone china, the maturing temperature of the body began to exceed that of the glaze, driving the need for high-fired biscuit ware and a low-fired glaze firing. The biscuit firing of ware became the norm in both industrial and studio ceramics. The difference with the latter being a low biscuit firing, with a higher temperature firing for the glaze and body to mature together, as with terracotta, stoneware and porcelain. A biscuit kiln can be packed efficiently by placing items inside one another to maximize capacity. Faulty ware with splits and cracks etc. can be rejected before glazing, so possibly three or more glaze firings could be obtained from one biscuit firing. More complex forms can be created and supported by calcined alumina or placed on setters for items that would not withstand the rigours of being glazed at the clay or low-fired biscuit stage. It is also true that some potters prefer this approach for its technical demands and aesthetic. More can be read on this topic in the book *Single Firing* by Fran Tristram.

TIP: You can glaze the pots when leather hard or dry them out completely. I would do this by putting them on a kiln that was firing overnight, then glaze them while still warm; I have found this results in less slaking.

Freshly thrown plates drying in the sun before biscuit firing. Photo © Kevin Millward

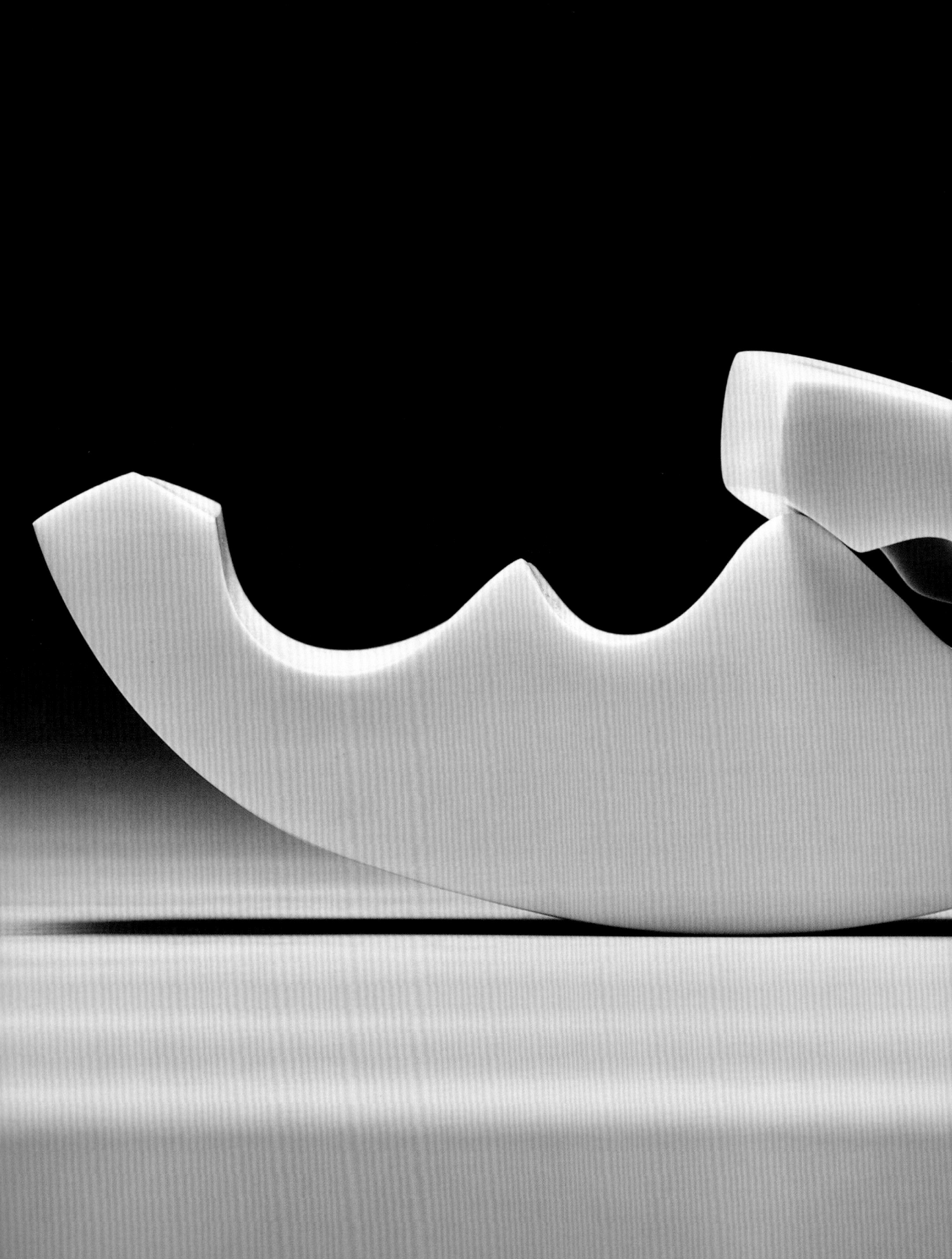

CHAPTER 15

POTTERS

The next chapter is about professional potters who are out there making high-quality ceramics in a wide range of styles and using varying firing techniques. How to they approach their practise? How do they justify the use of processes and materials that some might consider polluting? What do they do to lessen their impact on the environment? Maybe you could take a similar approach in the way you work, making your impact less detrimental to the environment. Take inspiration from their practise.

These are potters I know and trust. Their work is of the highest quality, they all respect, and are aware of, how their practise impacts the environment. I am not the arbiter of taste or style, but we should all try to be as good a potter as we can be and respect the planet we live on.

Angela Verdon bone china form. Photo © Brian Slater

Kiln packed with raw pots ready for firing. Photo © Lisa Hammond

LISA HAMMOND

Lisa is regarded by many as one of the leaders of the soda-firing revolution. Her work is heavily influenced by the Japanese aesthetic and her philosophy is not just based on the more environmentally friendly use of soda as opposed to salt (although this is contested by some), but by the unique quality it can bring to the surface of a pot.

Over many years Lisa has perfected the art of placing her pots to achieve the best possible results from the soda. Her work is also once fired, due to the size of her kilns and the type of burners she is using. It is imperative that the kiln does not heat up too fast, and this has led to the development of a system of smaller burners as well as the main burners, referred to as 'jockey burners' as they sit on top of the main burners. This enables the preheating of the kiln, preventing any explosions of the raw pots (non-biscuit fired). This is normally done overnight. The main burners are lit when the appropriate temperature is reached and there is no longer moisture left in the kiln. If you adopt this system and you are always not with the kiln, it is imperative that your burners have flame failure on them. This is a legal requirement. The aim of this technique is to get the biscuit firing done in the early stages until the you get to 995°C (1823°F). When the appropriate temperature is reached, you can progress to the reduction part of the firing and then continue to the high temperature firing. When the desired temperature is reached, and the relevant cone is down, Lisa commences the introduction of soda into her kilns. The soda is sprayed into the kiln chamber as a super saturated solution; this is achieved using boiling water to dissolve the soda. The amount of soda introduced is dictated by the size of the kiln and how many times the kiln has been used. Draw rings are used as well as cones to determine whether the pots have been exposed to enough soda and heat work. If all is well, the firing can be terminated. The kiln is sealed up and allowed to cool naturally.

Lisa fires her kiln on natural gas, as her studio stems from a time when it was viable to have this installed, a rare occurrence today. If she were to move studio, unless a gas supply was already present, the cost of installation would be prohibitive, necessitating the use of other fuel types.

Lisa checking results after soda firing.
Photo © Lisa Hammond

ANGELA VERDON

Since leaving The Royal College of Art, Angela has worked predominantly with bone china which she fires in electric kilns. All her work is unglazed; the whiteness and translucency are unsurpassed. Due to its lack of plasticity, the most practical way to use bone china is by slip casting traditionally, which can take only a matter of minutes; the process allows you to able to pick up every intricacy and nuance from the plaster mould.

Angela leaves hers in for up to 45 minutes. This enables her to manipulate the forms when removed from the moulds, while still wet in the core of the cast. This time is variable dependent on the size and shape of the forms. Once dried out, the forms are fettled and sponged to remove any minor imperfections, then biscuit fired to about 1000°C (1832°F). Although the piece is fragile at this stage, the surface can be refined using wet and dry sandpaper and diamond pads before being placed on, and supported by, kiln furniture. Calcined alumina acts as a setter, to enable controlled distortion during the firing, achieving vitrification at 1260°C (2300°F).

A glaze-like finish is achieved by polishing again with even finer wet and dry paper and a series of diamond pads, giving a white, super-smooth surface that pushes the boundaries of what is technically an unglazed surface, producing abstract forms through controlled heat work that captures movement and fluidity. Angela has a maximum of three firings a year as she is very aware of how her practice impacts the environment.

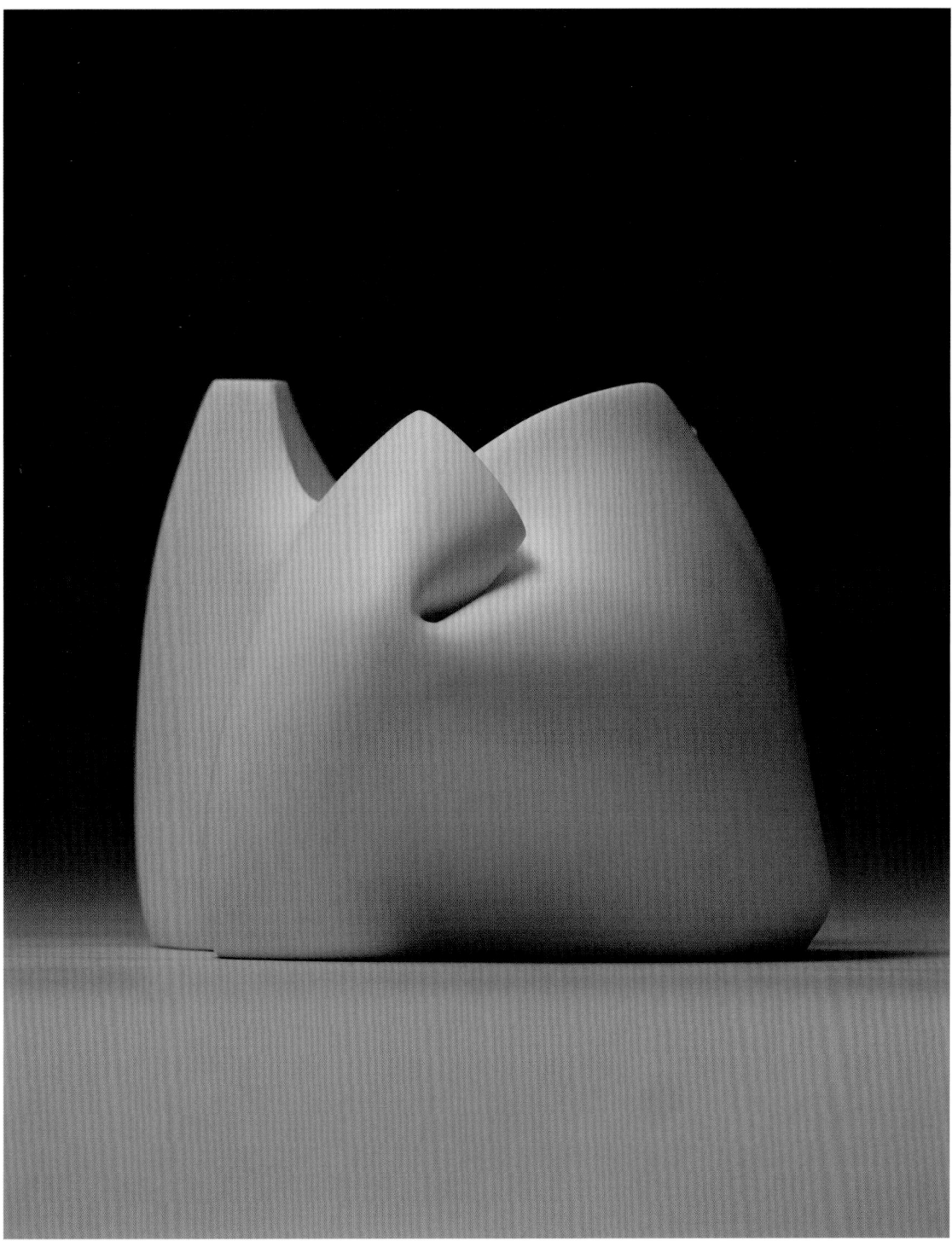

Bone china form. Photo © Brian Slater

Green polychrome glazed jug. Photo © Walter Keeler

WALTER KEELER

Walter's work is instantly recognisable as salt glazed. Ever aware of the environmental impact of this process, he has always endeavoured to produce work of the highest quality. His salt kiln is of a traditional construction in heavy fire brick to tolerate the repeated salt firings, backed up with a ceramic fibre blanket to improve thermal efficiency. He uses an oil burner system that burns efficiently and has a very low environmental impact.

When Walter started to make earthenware pots it came as a surprise to some, as it is very different to his salt-glazed work. Anyone familiar with the rise of ceramics in Staffordshire can spot the historic link that has inspired his forms and methods of glazing. Walter biscuit fires his cream earthenware body in an electric kiln at 1150°C (2102°F) to ensure strength and craze resistance. The lead-based glaze must be set up to be applied to biscuit ware with little or no porosity, and once dry the work is fired at about 1020°C (1868°F). For more information, see the chapter on methods of glaze application.

Left: Green polychrome glazed box. Above: Yellow polychrome glazed tea pot box. Photos © Walter Keeler

Slipped and glazed terracotta money box. Photo © Jessica Turrell

RUSSELL KINGSTON

&

JESSICA TURRELL

Lynmouth Pottery

Russell and Jess are two young potters who have chosen to work in terracotta, using slips and a low solubility lead glaze. The popularity of their work is reflected by the demand for the pots they produce. Most potters working in this field use electric kilns as their main method of firing – these kilns are oxidised and relatively low fired, with little wear and tear on the kiln and elements.

Makers often find that their orders outstrip the capacity of the kilns they are using due to the electricity supply. Many larger kilns would require three-phase electric power, which is not always available, or natural gas, the instillation cost of which is prohibitive. So, when the need for a larger kiln arises, what are the options: wood, oil, or gas? If you require a clean oxidising atmosphere your circumstances could well rule out wood and oil, so propane may be the way to go. Russell and Jess found themselves in the situation where the demand for their work exceeded what they could process through their electric kiln. The cost implications of purchasing a gas kiln of a large enough size to increase production was prohibitive, so, like many potters before them, they decided to build their own. The main considerations for them were the cost of firing, the fuel source, the size of kiln and construction materials used. Choosing their size of kiln was an important factor as if it was too big it could take too long to fill it, but if too small it would need to be fired more often, so could potentially affect their production and cash flow. They chose to use propane as their fuel, powering four burners and HTI 23 as the main insulating brick, held together with an angle iron frame.

Slipped trailed terracotta bottle.
Photo © Russell Kingstone

ROGER COCKRAM

Once Firing With Porcelain

It may surprise the reader, but in my experience and studio practice, I have found that the principles behind once firing and glazing apply to both stoneware and porcelain. My basic techniques of glazing – glazing the inside first, then the outside when bone dry, I perform in the same way for both materials.

Very briefly, I was lucky enough to be trained at Harrow College of Art in the mid 1970s, under a great potter, Mick Casson. The course was special in many ways; it was quite similar to a 'group apprenticeship', whereby the training was carried out by practising artists and potters such as Wally Keeler, Colin Pearson, Russell Collins and Mo Jupp, all of whom came in on a regular, part-time basis. This had the effect of augmenting our theory learning with real workshop practice, both in throwing, hand building, kiln building, as well as glazing and firing.

I was first taught to biscuit fire the work and devise suitable glazes accordingly, but Colin Pearson later said to me one day, "Right, Roger, now I'll show you how to never biscuit fire again." He never did biscuit firings – and since then, neither have I. I'm sure there are situations where it may be considered a good idea to do so (perhaps if using fine earthenware

Raw glazed jar in porcelain. Photo © Roger Cockram

for example). It's just that I've personally never found the need.

From the very beginning, I've always made some domestic ware using a stoneware body. However, for many years now, I have been using porcelain for my non-domestic, individual pieces.

Both my individual work and the domestic work are fired in the same mains gas kiln of about 0.76 m³ (27 ft³) to cone 10/11 in quite strong reduction. The glazes I make are modified from the 'norm', (I would call it 'put back') by the removal of china clay from the recipe and the insertion of ball clay.

Now a brief lesson here; in essence, ball clay recipes, used perhaps in a factory situation, would have shrunk off a biscuit pot before the firing. So, by using china clay, which hardly shrinks at all, at the glazing stage it was found that the two married together well. However, in a studio situation, by replacing the china clay with ball clay, the glaze and the unfired pot are able to shrink at the same rate and so fit together well. I could use up to 25% ball clay in some clear glaze recipes and more for matt glazes.

I also do two things for my base glaze recipe.

1. I use an additional 2-3% of bentonite in the glaze. This acts as an aid to shrinkage and fit.

2. My base celadon glaze uses more than one flux to make up a given percentage. For example, if the total flux in a recipe is 30%, then I usually use two or even three fluxes to make the same total percent. This has the advantage of giving a good melt in the celadon base glaze and increases the maturing range of the glaze. It also makes a good tight fit with the body.

The beauty of this assembly, which frankly came as a surprise to me, is that it seems to work for both the stoneware body and my porcelain. Sometimes, I make my own porcelain body in a clean plastic dustbin, using china clay, FFF feldspar, quartz and white bentonite. I mix this with water to a thick 'Devonshire cream' consistency, sometimes adding a fine molochite grog, to open it up a little, for throwing large pieces. It is then slowly dried back and kneaded for use. Alternatively, I buy the Audrey Blackman body recipe, which is excellent.

The glazing is done on the inside of the pot when it is quite firm but still damp, such as when one might use slip. Rather like the consistency of cheddar cheese left out for a day. I simply pour it in and out then leave the pot to dry slowly. The outside needs to be bone dry, when it can be dipped or glaze poured, as if the piece has been biscuit fired. In the past, I sometimes sprayed the piece, in stages, just to be careful.

Personally, I'm interested in the sense of depth and movement in the sea near my home and I seek to echo this on my work. So, I finally return the pot to the wheel before firing and brush on additional glazes, some layers of which are designed to flux more and move the upper layer and give that movement and sense of depth that I'm after. Recently, I have sometimes used gold lustre to suggest a hint of the setting sun on the surface. To achieve this, I apply resist to the surfaces in small areas before any outside glaze. Then after the main firing, I paint in the gold on the bare patches and re-fire the work to around 780°C (1436°F). Obviously, although the pot technically has no biscuit firing, it would be foolish not to heat the kiln gradually in the early stages. However, after 1000°C (1832°F), I start to reduce the kiln and then the firing proceeds from there, in my case with a prolonged last stage, to get a good interaction in my glaze layers, stopping when cone 11 starts to move. An hour with everything open follows. This sets the glazes and then it's time to close it all uptight and wait for everything to cool.

So there it is. If these techniques with porcelain are considered unusual and of interest, I hope it's been helpful to read about them.

The advantages are:
1. No chance of dust on the biscuit, producing crawling of the glaze.

2. A smoother single firing is certainly cheaper. I just make the piece and then fire it.

The disadvantages are:
1. Most of my work is thrown on the wheel, which must make glazing easier.

2. The single firing takes longer.

Galloway sphere. Photo © Matthew Blackley

MATTHEW BLAKELY

My work re-establishes the link between pottery and place, making pots entirely from rocks and clays that I have collected from individual locations around the country. These place-specific pieces are made entirely from materials that I have researched, collected and prepared myself. The glazes are blends of a variety of local rocks, and so are ceramic representations of the geology of those places. All the colours and textures are produced by the minerals present in these rocks.

Provenance is fundamentally important. Not only do I know where every ingredient comes from, but I have collected it (having gained the necessary permissions) and prepared it myself. Time becomes an inescapable part of the working process, in which there are no shortcuts. Every rock must be crushed, ground and sieved before it can have used. Hundreds of glaze tests must be carried out before I am able to achieve my results. I am very aware that pottery making relies on

the extraction of naturally occurring materials and requires energy to fire the pots, and I endeavour to work in ways that minimize my impact. I know the provenance of all the materials I use and collect them with respect for the places they come from and the people or organizations that own the land. None of these materials are toxic, iron being the only colourant I use.

Firing requires a significant amount of energy, which I try to minimize. I have built thermally efficient kilns with designs that allow me to fire without producing much smoke. I use local wood that would otherwise be burnt as waste, such as coppicing trimmings from farm and byway management, and branches and joints from fallen trees that are difficult to split. My firings are relatively short at 27 hours, though I am able to develop rich wood-fired surfaces as I am particularly considerate of the most appropriate clays, slips and glazes.

Edinburgh sphere. Photo © Matthew Blackley

DENNIS FARRELL

I have been making ceramics for over 50 years. Much of my early work was influenced by urban environments in the midst of sudden change. My current work responds to light, line, colour and texture, observed in rural and coastal landscapes.

Details in the landscape and weatherworn eroded structures are of particular interest, whilst atmosphere and sense of place are important elements.

Whilst I have always used photography and drawing to gather and develop ideas, I have more recently expressed ideas by directly painting with oil and acrylic mediums. As I'd always painted on clay, the move to painting seemed a natural step. However, this new world of colour and scale has at the same time been exciting and challenging.

Forms are produced by hand-built techniques using red earthenware blended with black stoneware crank clays. White engobe is applied to the clay surface sometimes, being drawn through when wet. Brushed, coloured engobe and underglaze colour are then worked together on the leather-hard clay surface like paint on canvas. Forms are brushed with semi-transparent glaze and fired to 1120°C (2048°F) in a 0.28 m^3 (10 ft^3) electric top-loading kiln partly fired on solar energy when generating power is possible.

Post firing, some surfaces are reworked using diamond pads.

'Windswept Shore'. Overleaf: 'Hidden Valley'. Photos © Dennis Farrell

Above: Gun metal tea pot. Overleaf: Gun metal bowl. Photos © Suleyman Saba

SULEYMAN SABA

The kiln firing methods that I use as a potter changed when I established a London-based studio in 1997. Up till 1996 I had worked alongside potters who used gas kilns for producing reduction-fired stoneware. I took the excitement of reduction firing for granted until the opportunity of setting up my own studio presented itself, after I moved to live in Camberwell, South London, in 1995.

The layout of the mid-terrace house where I live made it possible to develop a studio at street level. As the studio's proximity is close to a primary school, I felt it was impractical to build or install a gas kiln. Despite an unfounded wariness of electric kilns, I realised that adaptation was part of the course of a potter's working life. So, in 1997, I installed a reconditioned 0.28 m^3 (10 ft^3) Kilns & Furnaces front-loading, electric kiln (the largest that could run on a single-phase supply), and changed the electricity meter over to Economy 7, which the London Electricity Board did free of charge!

Fume extraction was resolved by siting the kiln within a passageway underneath the property. With the kiln being outdoors, I was concerned about the effects of damp upon the electrics during the cold months; but intermittent firings solved this problem.

I had been making stoneware since art college, and wanted to continue exploring its possibilities. There did not seem to be any point in trying to recreate reduction conditions by throwing in splinters of wood or mothballs during a firing, knowing this would reduce the working life of the kiln elements. I thought it would be an exciting challenge, as well as playing to the strengths of an oxidising atmosphere, to develop glazes that pushed materials in different directions, rather than relying upon the vagaries of reduction to determine the fired result. The first few firings were crammed with hundreds of glaze tests applied to three clays for comparative results: white stoneware, a darker stoneware and a porcelain. In addition to testing published recipes, I tried formulations of my own, including numerous line blends, some of which are still in use after 25 years. Other technical considerations began to emerge in learning to fire with electricity. I soon discovered that frequent high firing to cone 9 at 1280°C (2336°F) or above would tax the working life of the kiln elements. It took a bit of time to determine optimum firing temperatures to obtain glazes that would inform my work. As it turned out, some of the classic Chinese glazes that I had wanted to emulate – oil spot and iron reds – were possible to recreate under oxidising conditions, and not necessarily at the highest temperatures.

I finally felt that I could leave gas kilns behind. As the saying goes: "When one door closes, another opens."

One of the biggest challenges in using electric kilns is learning to understand the heat. You can't rely on temperature readings from an electric kiln controller to determine if glazes have reached maturity, because readings only indicate the ambient temperature within the kiln chamber, not the heat work upon materials. Similarly, custom-built electric kilns do not allow for removal of draw trials during a firing, in the way that bespoke salt/soda or wood kilns do. My earlier experiences of gas firing taught me about the value of pyrometric cones to gauge the actual heat work upon clay and glazes, and I continue to use pyrometric cones to monitor the length of soak at the peak of the firing cycle.

This enables me to achieve accurate and repeatable results. Because the kiln is outdoors, weather conditions can encourage fast cooling and unforeseen problems such as dunting during the cold months. I add soaking times into a glost firing program (after the top soak) at incrementally lower temperatures to replicate the cooling of an indoor kiln. The kiln door has two small bung holes that coincide with the upper and bottom levels of the chamber. There can be a temperature difference of

30-40°C (86-104°F) between the bottom and top. I regulate heat distribution by staggering the arrangement of shelves between the rear and front portions of the kiln chamber, as one would in a gas kiln, to encourage the path of a flame from burner port to exit flue. If firing between December and March, I keep the kiln shelves indoors until required, to minimise the effects of frost and thermal shock to the furniture. A more openly filled chamber, rather than one that is densely packed, encourages the kiln to reach the required top temperature more quickly. I position glazed work throughout the kiln to take advantage of temperature variables and monitor the bending of cones from the lower bung hole. It's ironic in how the interplay of working with gas kilns has informed how I fire an electric kiln!

Postscript: from about 2018, I started to use mid-temperature – cone 5/6, or 1200°C (2192°F) – firing techniques to reduce energy consumption for the kiln. Although I still occasionally fire to cone 8/9, I have enjoyed developing new glazes at cone 5/6 for both tableware and sculptural work. I don't fire the kiln as frequently as in earlier years, so I always try to test new combinations of materials alongside stock and commissioned work.

Potters are a wily bunch and creative, with the availability of limited resources at their fingertips. I believe that mid-temperature firing can go a long way to reducing energy consumption from the National Grid, without compromising artistic ideas.

ADAM FREW

I use Mason stains for my bright colours, mixing them into a slip which is only china clay.

I use cobalt oxide a lot when decorating my work and add a small percentage of iron to take the intensity out of the blue a little.

I play with the thickness of the slip and can get different effects at the leather-hard and bone-dry stages. I sometimes like to apply my slip like watercolour.

I also use onglaze like a slip and have started making my own onglazes with around 80% of feldspar and 20% china clay which gives the colour I'm using more of a shine. When I apply this at the leather-hard stage, and it mixes with the porcelain, I get more of a satin finish.

All my decoration is done before biscuit firing. I also make my own ceramic crayon which is the cobalt and china clay slip dried out and rolled into a crayon shape.

I glaze fire my pots in a Rohde Gas kiln that is designed to be very fuel efficient for reduction firing, as I use a celadon glaze on most of my work. This necessitates the use of uses propane as natural gas is not available.

Cobalt decorated globular vase. Photo © Adam Frew

Slip decorated, once fired terracotta bowls. Photo © Josie Walter

JOSIE WALTER

Philosophy/approach

My marks are continually evolving. I am interested in contrasts; sharp lines, crayon scribbles, brush marks, sponged back sections. Working on green ware allows more movement in the material, more organic flow of the marks. It is this ongoing investigation that invigorates my making.

As with many potters, my trajectory into working with clay took a rather roundabout route. After a degree in Anthropology at UCL from 1969-1972, I spent some time studying engravings in a remote cave called Los Hornos de la Pena in the Cantabrian mountains in Spain.

On my return I decided to take a PGCE as a backup to cave studies, followed by teaching in secondary schools for the next four years. In my first post I was used a bit like cement on the timetable, filling in where there were gaps. I taught English, Maths, Complimentary Studies and, horror of horrors, one lesson of pottery! The grammar school I had gone to considered art irrelevant, so I had no art experience whatsoever apart from one term at age 11, when I produced a very dull painting of a geranium.

I decided my best plan was to join a pottery evening class. It was love at first sight. The evening class was followed by summer schools at Loughborough Art School and the finally a Studio Pottery course at Chesterfield College of Art and Technology from 1976-79. The course was taught by Harry Hibberd, Trevor Nicklin and Geoffrey Fuller. Geoff was to be the most influential in the long term, although I was not aware of this at the time. His playfulness and humour, his love of history, the interaction of slip and soft clay, momentum wheels and once firing, has always permeated my approach to making.

In 1978 I set up The Courtyard in Matlock with John and Judy Gibson and friends from teaching. We opened a fabric shop, a wholefood store, a restaurant and the pottery workshop and shop. We left college in June and moved straight into our new premises, but I soon realised that I needed more experience. An ad in Ceramic Review for a thrower in France caught my eye. The following April I was at Le Poterie du Don, in the Auvergne, making salt-glazed stoneware, learning about pots and food and grappling with my French. It was an extremely instructive time, and I came back full of ideas. In 1986 I started working part time at Derby College, soon to become University of Derby. I began in the pottery workshop, but my love of history soon led me into academic teaching, not only in ceramics but also in fashion, textiles, illustration and finally animation. The students were a joy and the different disciplines a rich area of resources for me.

The University also encouraged research, providing funding for me to take an MA at Staffordshire University in the History of Ceramics which consequently led to my book *Pots in the Kitchen*, published by Crowood in 2002.

I left my three days teaching in 2014 to work full time in my workshop at home. We don't use the word 'retirement' in our house, but at the same time, caring for my mother and then grandchildren took over my days and lockdown was actually a bit of a haven. Not able to go anywhere, time was luxurious in the workshop. I took some drawing courses, hugely beneficial to expanding the imagery on my pots and I worked away to develop new ideas.

Unable to go to fairs and with galleries closing their doors, I started an online shop. This was a steep learning curve but has been great. In fact, I'm now rather reluctant to race round the country to stand in a marquee in my wellies when sales can be done from home. I miss the socialising of course, so some fairs will be on the cards, but maybe just the local ones!

The workshop is next to the house, just up a few steps, so I still have the feeling of 'going out to work'. The first job is to light the wood stove, then check the pots in progress. All my pottery is made in red earthenware clay, thrown on a momentum wheel. I also make some slab pots such as butter slabs, spoon rests and square and rectangular trays, using a soft clay technique, beating the corners together with a wooden batten, rather than constructing pots using hard slabs and slip. All the thrown pots are also made with a fairly soft clay that keeps turning to a minimum, hence very little waste clay or reclaim. The pots are decorated with a layer of white slip then with images of hens, vegetables, fruits and animals such as pigs, deer, goats and cats. After decorating, the pots are raw glazed and left to dry. This type of glaze is rather like a slip – that is, clay and water – but with an added ingredient, a frit which when fired makes a shiny surface on the pots.

When the pots are dry, they are decorated with coloured glazes, packed into an electric kiln and fired to around 1100°C (2012°F). I just have this one firing, that is no biscuit firing. I have just had a kiln controller fitted so I no longer need to nip out to the pottery in my nightie to turn the kiln up. The kiln goes on at 10 p.m. and off at 8.30 a.m., making good use of my cheaper night saver meter and I also get a good night's sleep!

The interest in sustainability in practice is very much at the forefront of thinking today. It is important to be hyper-aware of the impact our work has on the environment and how to have a greener practice. Using a momentum wheel, once firing, economising on electricity, minimising waste and working from home, so reducing car use, are all heading in the right direction. Making useful pots to last long-term has also been one of my aims. I like nothing better than a customer coming up to me and saying, "You know that mug you made 40 years ago…could you make me another?"

Overleaf: Slip decorated, press moulded dish. Photo © Josie Walter

KAREN ATHERLEY

Over my ceramics career so far, I have had four top-loading electric kilns.

The first one I bought in 1983 when I finished at Camberwell and was setting up a South London studio in Sumner Road, Peckham. Together, Philip Vain and I bought the equipment, and other potters paid for their space.

We were taught about packing kilns, loading and firing them by our great technician Denis at Camberwell, where we mainly fired in large, front-loading electric and gas kilns as well as raku and salt-fired kilns in the yard.

Although this was fun and great experience, the work we were doing didn't need that sort of kiln. In 1983 the American hobby craft market was taking off and new top-loading kilns were being made and exported. Potterycrafts was updating its products and catalogue, so we got to know Ken Shelton at this time.

I bought a Amico Gold top-loading kiln; it was quite large and could fit a variety of work in it, and it was a lot cheaper to buy than the more traditional kiln we had used. The beauty of these kilns was they had a kiln sitter; a three-pronged unit where you put your cone and when it reached the desired temperature, it would switch itself off. You could also program how fast you wanted it to go, which was also novel as it was a lot quicker than the big front loaders. There was no need to sit and watch it, although we did monitor it!

The work that we were doing in the studio also influenced our decisions as we were mainly firing to earthenware

Underglaze decorated jug. Photo @ Paul Lapsley

Underglaze decorated flower brick. Photo @ Paul Lapsley

temperatures, but it could have gone to stoneware also. When I left to go and live in Peterborough and set up my own workshop, I sold my Amico Gold top loader and went for a British top-loading kiln, which I bought from Potterycrafts. This one had a different kiln controller, a Stafford; this would fire to the temperature you had programmed it to and switch itself off. For me and the work I do it works well, it's not complicated and it does what it says on the tin. They were advertised as hobby kilns but I think are a great addition to a very varied way of producing pots. In 1996 I moved to America with my husband's work, where I set up another studio and bought a Skutt Kiln. This top loader was even more up to date and the controls were all part of the kiln, not a separate box connected on the side. The American market was very innovative for small-scale production and was producing lots of helpful glazes and coloured slips that were already made up for the pottery market. These were also exported to the UK at this time. I think it opened a door for other kinds of work to be produced, without having to have a vast amount of technical knowledge, but for myself it gave me a great palette of colour and a safe knowledge that what I put in the kin would come out OK (on the whole – we are talking ceramics, where anything can happen!). This went for firing the Skutt kiln also, as it has programs for all sorts of temperatures.

On my return to the UK, I brought the Skutt kiln back and at this time Ken Shelton was at Cromarty testing the Skutt kilns for distribution in the UK. I wasn't sure at the time what the reaction would be to them, as I felt they were a bit too basic but after years of service it turned out to be a great kiln.

At this point, living in a village in Lincolnshire with not many amenities, and oil being our way of heating the house, we looked into having solar panels fitted on the roof. At the time this was part of a government scheme where they rented your roof space, and you got a certain amount of free electricity. This helped enormously with bills at the time, and I would recommend anyone currently firing electric kilns to look into them, as they can save you money in the long run. Like anything, it's the initial outlay that can be off-putting, although there to seem to be a wider range of different prices available, so do consider them when thinking about setting up a workshop.

Scrap wood before cutting. Photo © Ben Brierley

BEN BRIERLEY

Cross-draft wood-fired kiln.

I wasn't initially drawn to firing with wood 25 years ago, for several reasons. Although awareness and understanding of our environmental impacts and consequences have matured, and materials and kiln technologies have evolved, these reasons still ring true today.

Wood, as a firing material in the UK, is a relatively renewable and sustainable resource. I have always used waste material from wood yards involved in the production of gates, fences, building materials etc. for agricultural and domestic use. When I approached my first wood yard, their scrap wood was burnt on site, creating a lot of smoke which the workers at the yard were almost permanently swathed in. When I offered to take the offcuts away and use them creatively, the yard was very supportive, as it improved their working conditions and provided me with fuel for my kiln. In return for the wood, I supplied the yard with mugs fired using their waste material.

Suppliers of fuel have changed over the years, but the use of a waste product has not. The relationship between myself and the folks working in the yards has been one that I have cherished, as we are all in some way part of the finished objects. The wood yards I have used usually source their timber from forestry commission auctions, buying in complete tree trunks which are then edged of their bark in preparation for being turned into boards or planks. It is these edgings which I use. The barked outer layers of the trunk are the part of the tree that contain most of the minerals drawn from the ground in which it grows. These will be liberated on burning and work with the silica, iron and feldspars in the clays to create complex, rich surfaces.

Although predominantly soft wood edgings, I additionally use a few bundles of hardwood slats each firing, taken again as a waste product of creating tongue and groove hardwood boards. I used to receive free timber from several tree surgeons in the area for many years until wood burners became fashionable and it became viable for the businesses to spend the time processing the waste for wood burner fuel to sell. Surfaces that a wood flame can produce on ceramic work, when the chemical elements contained within it react with clay, create patinas and subtle tones that cannot be achieved in any other way. Elemental surfaces, sometimes challenging surfaces, that speak of the process, the history and integrity of human engagement. We can set up a series of potential outcomes by the way we pack the kiln, the materials we use, the wood we use, the way we choose to fire the kiln. Ultimately though, we have little control over the way the clay will respond to flame movements instigated by these external decisions. As a maker I find this exciting; that we can relinquish control and put our trust in the choice of materials and process.

Firing with wood requires a continuous engagement with the process to finish the work. It involves the preparation of the fuel. In my cutting approximately four tons of 3-4 m lengths of slab wood into roughly half meter lengths and then splitting and stacking to dry. The processes involved with firing a wood-fired, cross-draught kiln for around 80 hours, has always been compelling for me as a maker. The sourcing of the wood, cutting it to the right length, splitting it so that it seasons, stacking. All the preparation is part of the ritual of firing. Once the kiln is packed, a process comparable to trying to anticipate the reaction of water when placing pebbles in a stream, the work is committed to all the variables of the extended firing.

It is my job to listen to this structure and respond by controlling air intake or chimney draw to keep the kiln climbing or instigating atmospheric changes within the firing chamber by adjusting stoking patterns. I have to attempt to grasp and understand elements of the chemistry involved to make educated guesses on what ceramic materials may be doing, working with experience and hunch. Ultimately the resulting work will be a product of all these activities.

Group of wood fired bottles. Photo © Ben Brierley

JANE WHITE

Jane White has made full use of the environment where she lives to practise the pit-firing technique to fire her ceramics. Jane and her husband farm in an isolated valley in the Chiltern Hills, surrounded by woodland, with the added advantage of having access to farm machinery for collecting up the wood used for the firings. She is also able to use the farm digger to initially construct the large deep pit.

"It seemed the ideal choice of firing for my situation, with unlimited access to wood from the fallen branches, and living in an isolated location, where having a large fire wouldn't create a nuisance to neighbours. The wood I use is mainly beech and ash, and as the firings are done in the summer months, the dead wood is dry, so doesn't need to be stored before being used. I source the sawdust used to line the bottom of the pit from a furniture maker, who has a workshop at my daughter's farm. He likes to make up 'special mixes' for me, usually a mixture of hardwoods.

To ignite the fire, and ensure an even burn, waste straw is woven through the wood loaded in the pit, so the fire burns with a hot, clear flame rather than smoking the work. I also use lots of redundant heavy metal sheets that I found around the farm to cover the pit towards the end of the firing, which creates a reduction atmosphere, and slows down the cooling, which further reduces losses. The organic materials, dried fruit skins, stones and coffee

grounds that I use in the pit firings are all gathered over the year and dried in the bottom oven of the range cooker, which is permanently on a low temperature as a consequence of being the main source for heat, cooking and hot water in the farmhouse.

Whenever I visit friends who live on the coast, I collect the dry, drift seaweed washed up on the foreshore and found at the fringes of beaches and I store it in old bread crates for use in the firings. Before gathering seaweed, it is always best to check first whether you need permission from either Natural England, the council, or a private landowner. You can collect up to 20 litres of unattached drift seaweed per day for personal use, although this may vary according to the local council. Driftwood found on beaches is also a valuable resource, as it is embedded with sea salt, so will give lovely orange colours. Although it is possible to raw fire the ceramics in a pit fire, it is a very risky option, as the temperature rise is fairly rapid and uncontrollable. I invest a lot of time in each piece, so I prefer to low fire the work to 890°C (1634°F) in my electric kiln before pit firing, which ensures a greater survival rate. All my pieces are made using Ashraf Hanna clay, burnished, and then three layers of terra sigillata are applied to the surface of the bone-dry work, which is then immediately polished with a soft cloth to create a brilliant glaze-like shine. Terra sigillata means 'sealed earth', and is a very fine, smooth, virtually waterproof, clay slip. Applying it as a surface treatment removes the necessity to have lots of chemicals in my studio as the work has its own 'natural' glaze. About 50 pieces of ceramics, approximately four months' work, can be fired in the pit, thus making full use of all the resources."

Previous page: Pit-fired vase
Left: Pit- fired bowl. Photos © Jane White

'Salvage Series: Buller'. Overleaf: 'Salvage Series: Trow'. Photos © Guy Evans

NEIL BROWNSWORD

Salvage Series

With the industrialisation of ceramics during the eighteenth century, systems of segregated labour brought about a phenomenal concentration of specialist skills and knowledge to specific regions of North Staffordshire.

As much of this anonymous dexterity remains largely overlooked, a compulsion to illuminate the actions implicit within specific divisions of labour continues to be an ongoing artistic concern. These works aim to emphasise the 'human' element of manufacture by referencing or incorporating within its fabric, discarded clay detritus salvaged from the factory production line.

Rejected wares, turnings, and spares from castings, that speak of innate judgements essential to the success of outcomes, are preserved and aestheticized through firing. These momentary imprints of subliminal action retain a vigour and expression alien to mass production, where standardisation has rendered evidence of human contact an imperfection.

To further reverberate the daily routine of a rapidly disappearing culture of labour, artefacts recovered from redundant factories – sponges, ware boxes and plate packages – are soaked or lined with clay and fired to making their memory permanent. Likewise, obsolete manufacturing technologies unearthed from the shraff-lined foundations of houses in Stoke, are appropriated directly into structures to cite and connect with the skilled endeavours of the past.

PHEOBE
CUMMINGS

Phoebe is a ceramic artist working in clay, but in many ways, you could say she is working in a most environmentally friendly way in that she does not fire work.

"In terms of thinking about working with raw clay I would say I was initially driven by a more personal sustainability: I couldn't afford to have a studio, kiln or tie material up in only one object.

However, that ephemerality and shifting nature of the work is also embedded deeply in what my work is.

It is not about a static object or possession, but focused on experience, and a sensory encounter with sculpture. I am interested in atmosphere, humidity and absence as much as I am interested in solid, physical material. I think about how we share breath with clay sculpture."

'Life and Death'. Photo © Roberto Salamone

Coffee cup and saucer. Photo © Glen Stoker

NATALIA
KASPRZYCKA

My involvement with found materials comes from an interest in the relationship between ceramics and place. Once the ordinary reality of a ceramics practice, with potteries traditionally located close to the sources of their raw materials, it's become diluted by the realities of modern life and the rise of ceramics as an urban practice. What ceramics materials can tell us about our local environment, its geology and transformation by humans is all the more poignant in this age of globalised production where not only the user but also the maker is far removed from the where our everyday objects come from.

Using local materials can be seen as sustainable in many ways. The maker doesn't need to travel far for their materials which significantly decreases the carbon footprint of their work.

Waste materials like shards, slags and some aggregates can be put to good use, giving what is considered waste another chance to shine. Perhaps most importantly though, finding, gathering, processing and testing found materials, especially in larger quantities, is a laborious and time intensive process, giving a sort of humility and respect for the Earth's resources and fellow manufactures all over the globe to develop, pushing one to carefully consider how and what one makes, buys and throws away.

Finding and understanding how to use found materials requires curiosity and a bit of research; after a while these just transform into an ability to spot potential anywhere. A basic understanding of geology and glaze chemistry is useful, but it develops though experimentation anyway. My usual process starts with a walk and looking at geological maps of the area (British Geological Survey has a fantastic selection of paper and online maps). Then my research moves to the human history of the area; was there an industry here or any other human activity which could be responsible for what can be found on site? Sometimes that process happens in reverse and what is present in a place can tell you its story. I collect samples of whatever I deem useful or intriguing.

PREPARING IRON-SMELTING SLAG AS A GLAZE INGREDIENT

The landscape of Stoke-on-Trent, where my studio is based, is an outcome of both its rich natural reserves and the history of ceramics, coal and steel industries. Pot banks and slag heaps are a familiar sight. For several years I've been investigating the properties and potential of materials found on one particular site, which between 1830 and 2005 was Shelton Bar steelworks, which in its prime spanned 400 acres. One material of special interest has been iron smelting slag, an obsidian-like waste product present there in large quantities. Iron smelting slag is the result of heating up iron ore (usually haematite) to separate the metal from other components of the rock, mostly silica. When the iron is collected, the molten silica with other impurities and metal oxides added sometimes as fluxes, are poured out, usually into a hole dug in the ground. They break apart into chunks of various shapes, sizes and colours upon rapid cooling. Being predominantly silica, this human-made rock can fulfil the same chemical function as quartz or flint in a glaze or slip. The surface effects created by metal oxides and other impurities will vary from batch to batch. To be usable in glazes and slips, slag needs to be ground to a very fine powder. I do this in two stages, first with a club hammer and then in a large porcelain pestle and mortar. I process slag wearing a respirator, good goggles and gloves, especially in the early stages; structurally the material is like glass and when you start breaking it apart it forms very sharp shards so great care needs to be taken. I prepare my working area which is a steel plate with a border of bricks and start bashing larger chunks with a club hummer to break it up into smaller pieces. These are then ground finely in a deep porcelain pestle and mortar. I grind slag down until the powder is fine enough to pass through a 100s sieve. Any coarser will result in bentonite additions to the glaze, which produce unfavourable results. The material can then be experimented with in slips and glazes as a substitute for other sources of silica.

IMPORTANT TIPS

When collecting and processing materials, especially post-industrial waste, you should wear PPE. Their composition is uncertain and could be harmful. Good respirators are essential during the crushing and grinding, which is best done somewhere with a good airflow – outside or indoors with extraction. No ceramics made with rocks or post-industrial waste should be considered food-safe unless their chemical composition is tested. Lucideon is a Stoke-on-Trent company which provides the service of testing the food safety of glazes.

Industrial waste and glaze trials. Photo © Natalia Kasprzycka

SHARON LEA

Sharon is a recent graduate from Clay College who decided she would set up her studio on the family farm. Her passion for reduction-fired stoneware and porcelain meant an electric kiln would only suffice for biscuit firing, meaning she would need to build her own kiln.

With advice from myself and Brian Dickinson it was decided to build a propane-fired kiln. The size of the kiln required the purchase of brand new HTI 23 bricks, and I advised her on how much they would cost and where to purchase them from. It was almost as if it was meant to be, as a large quantity of bricks became available from a TV production company I had been working with, who are well aware of their carbon footprint and will endeavour to recycle, re-use or sell on items left over from a production.

A perfect example of recycling. Sharon also takes clay she has found on their farm and uses it to formulate a glaze, giving a unique and individual quality to her work.

Previous page: Porcelain beaker. Left: Porcelain bowl. Above: Kiln. Photos © Sharon Lee.

Botanical slab-built form. Photo © Phil Wilkins

JACQUI ATKIN

Working History & Practice

Some 30-odd years ago, at the beginning of my potting career, I was very excited by the work of such makers as Magdalene Odundo and Gabrielle Coch whose forms were burnished and smoke fired. I loved the tactile surfaces of their coil-built forms, the amazing sheen they achieved through burnishing and the way in which the smoke visually enhanced the forms and added to their sensuality. I felt this was the technique for me, because apart from a kiln to bisque fire my work in, the smoke firing would be easy to achieve in a bin in my back garden and would be much more cost effective than expensive glaze firings.

However, I lived in a suburb of a town and hadn't quite reckoned on how much smoke the firings would make.

Sawdust can burn in a really acrid way and it upset the neighbours! I had to be careful about how and when I fired, which added a huge level of anxiety to the process. After some time of firing in this way I discovered the ceramics of Jane Perryman. Her work was still smoke fired but incorporated pattern on the surfaces. Having been a lifelong lover of surface design, this seemed the next logical way to develop my own work. I was lucky enough to do a workshop with Jane in the early nineties, where I learned you can fire with newspaper which burns quickly and has a less negative impact on the environment. The technique involved the same burnished surface vessels but introduced paper tape as the pattern decoration and a crank slip to act as a resist. The principle was to fire the work sufficiently for the paper design to burn away under the resist, leaving the pattern marks on the surface along with marbled effects from the smoke that seeps through surface cracks in the resist.

I worked in this way for many years, developing ever more sophisticated designs for my surfaces, and using only newspaper to fire the work, but it was labour intensive and often hit and miss. Sometimes a piece would have to be decorated and fired three or four times to achieve the depth of finish I desired which was no good if I had deadlines to meet. I began to research ways in which I might be able to achieve similar effects but firing in a faster way and, inspired at that time by the wonderful work of Dave Roberts, I began to experiment with resist raku techniques.

The challenge for me was to make the work entirely individual from other resist raku firers, there being many excellent makers out there using this technique. I experimented with resists on resists to achieve the effects I wanted and to introduce pattern other than just line repeats. In the course of experimenting, I varied from the fully burnished surface and tried adding texture as a contrast around the pattern. These forms were raku fired without resists for the wonderful depth and quality of black that only raku firing can achieve. During this time, I also dabbled in pit firing and loved the excitement of it and amazing atmospheric surfaces that could be achieved. But again, it is labour intensive and most safely carried out in the company of others, which made it impractical for me.

Of course, raku firing may be quicker, but it also involves a lot of smoke, which was less of an issue by that time because I had moved to the country where no one would be affected by the fumes. But no matter where I have lived, the issue has always been of niggling concern to me. However, my work progressed using this method of firing for several more years, having been liberated from the stress of its immediate effects on neighbours.

There was always one big issue in firing this way for me however, because I am seriously claustrophobic, so I did not wear breathing apparatus to fire and inevitably breathed in the smoke on occasion, despite my best efforts not to. I have never smoked a cigarette in

Above: Coiled form underglaze decoration. Overleaf: Tall vase, smoke fired. Photos © Phil Wilkins

my life but after having an operation which required a general anaesthetic, and having told the doctors I did not smoke, my lungs were foaming, and they didn't believe me. It was frightening. I realised that although I may not smoke cigarettes, I had been doing worse for years and that this was a warning that things had to change.

In an attempt to cheer me up my kids said, "Well, at least you won't smell like a smoked sausage anymore!" I hadn't realised I did – they could have told me sooner!

The challenge next was to find a method of working that replicated in some way the black surfaces achieved by the raku process, which I managed to do using velvet underglaze, but what was most exciting was that these also came in a huge range of colours which allowed me to introduce another dramatic element into the work. It was really liberating to be released from the extreme firing process which I realised, on reflection, had actually been making me feel quite unwell for several years. Added to this, my love of surface design could now be fully realised, and I could become a colourist.

At first, I stuck with the graphic contrast of black and colour but gradually began to expand the surface design and move away from black as the main focus. I searched for new ways to create designs, incorporating many textures and building up elaborate surfaces which reflect my love of birds, textiles and mid-century design. I made a very conscious decision to fire my work at relatively low temperatures, in part to achieve the colour effects I wanted, but also to keep costs down, along with wear and tear on the kiln.

Over the course of my career I have learned that there is a certain snobbishness attached to the way makers fire their work, and using electricity seems to be the least favoured, but it may well be the only choice in the not-too-distant future, and it arguably has the least impact on the environment. Moreover, when makers are prepared to experiment, there are few effects that can't be replicated using this method of firing, as I have learned over the course of the past six years as project editor for Clay Craft, a magazine for makers of mixed experience. This has involved me making work to fire to all temperatures to demonstrate the projects, but I try not to go above cone 6 now. With recent glaze developments and the introduction of safe, reliable, ready-prepared glazes, makers can go a long way to mitigating their effect on the planet and surely, the impact we have on our world should always be at the forefront of our practice.

I continue to unashamedly embrace the use of elaborate surface design and colour in my own work, having reached that wonderful and liberating stage in life, realising that time is precious and we should do what pleases us most and gives us joy.

MARK DALLY

Firing on Economy 7

Obviously, due to the massive increase in electricity over the last year, I have been looking at the way I fire and at my traditional (non-smart) meter, taking readings for Economy 7 and normal day rate and working out the costings far more precisely. The good thing about Economy 7 measuring is that all the lights are off and everyone is generally asleep.

When I'm not rushed, I do like to pack the kiln and fire it, so the peak (highest temperature soak) is firmly in the last hour of the Economy 7 cycle. There will always be some overlap as every firing is slightly different, especially if a large piece benefits with a cooling cycle to prevent dunting. My electricity supply is the domestic supply into the house, so unless I split the supply at great cost and have a separate business/pottery meter, there is always going to be an unknown element to assessing costs anyway. The heat from the kiln in winter is much appreciated by the family.

I make black and white functional domestic slipware using a combination of traditional craft and modern industrial ceramic techniques, in a contemporary take on Staffordshire slipware. Currently I am enjoying applying the studio processes I've refined in making my tableware and large functional work, to develop my sculptural one-off pieces.

I'm developing slab-built, stacked and highly decorated sculptures that are

Above: 'Trumpets'. Overleaf left: 'Boxer'; right: Large platter. Photos © Mark Dally

dependent on gravity to interlock. This making method offers me opportunities to create carved cutaways and pierced elements to increase transparency and introduce layers of visual depth. Additionally, I'm exploring and exploiting the differing reflective qualities of glazed and lustred surfaces. My recent sculpture series references contrasting themes of nature and technology, exploring shifting attitudes to cultural notions of worth and importance. Although my work starts with discarded, mass-produced everyday objects, as a ceramic artist I aim to apply refined workmanship and elaborately layered surfaces to imply value and legitimacy through transformation. I start by making plasterwork moulds of found objects. I use high-fired (1070°C (1958°F)) white earthenware to slip cast, wheel form and extrude my one-off tableware and large functional forms. I construct

each sculpture from a combination of slip cast, slab built, hand built, wheel-formed, sprigged and extruded earthenware elements which I join when leather hard. I layer the green ware surfaces with my characteristic black-and-white-stained brushed, stencilled, slip trailed and dripped slips.

I often use Adobe Photoshop to digitally manipulate new motifs for my hand-cut, paper-resist stencils. I am introducing selected raku-fired elements into my newest stacked sculptures, to contrast their rough smoked finish with the fine, high-gloss appearance of my dipped clear glaze, fired at 1150°C (2102°F). In a third firing at 780°C (1436°F) I apply liquid bright gold and platinum lustres, often to highlight a feature slip cast from discarded tubing or a cast-off plastic toy.

Above: Cubes. Overleaf: Chard vessel. Photos © David Binns

DAVID BINNS

Recycling Ceramic Waste

Having originally gone to university to study woodwork and cabinet making, David Binns found himself increasingly drawn to working with clay. However, his interest in wood has continued to inform his ceramic work throughout his career. During the early 2000s, Binns' work involved making large sculptural forms that relied heavily on the intrinsic visual appearance of the clay body. The aesthetic of each piece was provided by varying types and sizes of grogs (aggregates) he added to the base clay body – usually porcelain.

The aggregates he added to the clay body were either sourced from the refractory industry or he made them by staining clay, pre-firing it and crushing it, before adding to the clay body. After each piece was fired, it was ground and polished, to reveal the aggregate inclusions – much as a piece of stone can be polished to reveal its true beauty. Whilst being ceramic, this body of work had many parallels with his original interest in wood. Firstly, what was seen on the surface of a piece reflected exactly what was inside the material, much

akin to wood. And secondly, shaping and finishing often involved cutting and polishing, rather than relying on an applied skin of decorative embellishment or glaze.

As the project evolved, two factors swung Binns' work in a new and more sustainable direction. His desire to further expand the aesthetic possibilities of his work drove him to experiment with an increasingly wide range of aggregate additions. One avenue he began exploring involved adding increasingly large amounts of crushed glass cullet. Binns was also becoming increasingly conscious of the impact humans were having on the environment, the detrimental consequences associated with the quarrying and mining of virgin raw materials and the huge volumes of material sent to landfill. Considering these issues led him to start exploring alternative sources of raw materials. He initially experimented with small amounts of used glass bottles, broken tableware, which he crushed by hand, and other discarded ceramic waste from his university ceramics department.

As the project developed, it became necessary to find larger, more sustainable and reliable sources of recycled material. A local recycling company provided varying types of glass from recycled CRT TV screens and bottle banks. And a large sanitary ware manufacturer provided a reliable source of fired vitrified china from seconds quality sinks and toilets, which was otherwise destined for landfill. Adding crushed granular glass to a clay body was inevitably problematic, as the glass melts at a considerably lower temperature than that required to mature the clay body it was mixed into. Whilst increasingly large amount of glass additions gave increasingly interesting visual and textural effects, it also created an increasing technical problem. The solution Binns came up with was to fire the mix of glass and ceramic waste in moulds, to contain the flow of molten glass. In turn, using a mould to contain the mix when being fired meant he could explore adding even larger volumes of glass, mixed with the vitrified china, to the point where he eliminated the need for any plastic clay. His work was now made entirely from recycled, crushed waste glass and ceramic materials. To the base mix, Binns would then add other more interesting fragments of fired ceramic material, to further enhance the aesthetic appearance of his sculptural work. He found he could embed large fragments of broken tableware, which when polished back, gave the appearance of fossils embedded in the matrix of fired recycled glass and ceramic waste. Through embedding fragments of historical pottery and recycled material from a particular location, each piece of work held a narrative about the time or place of the shards it was made from.

The process Binns had developed of fusing blends of recycled waste glass and ceramics led to invitations to participate in two interesting projects. The first was a project to celebrate the important relationship between the UK canal network and the ceramic industry. Binns used discarded industrial ceramic shards dredged from the canals in Stoke-on-Trent, combined with historical pottery fragments found on the foreshore of the River Thames in London. Using this material combined with recycled waste glass, he cast a large boat-like form that symbolised the importance of the canals to the emerging ceramic industry. A second project involved an invitation to Mashiko, the pottery town in Japan, famous for being the home of Shoji Hamada. The tragic earthquake of 2011 had caused considerable damage to several historic kilns in the town, as well as creating huge volumes of broken pottery. The local ceramic association had heard about Binns' interest and knowledge of recycling waste ceramic materials, so they invited him to run a series of workshops, demonstrating his recycling process. This eventually led to the making of two signposts to celebrate the 100th anniversary of the connection between the Leach Pottery in St Ives, Cornwall, and Mashiko in Japan. Fragments of pottery and kilns from both towns were cast using Binns' unique ceramic recycling process. The signposts are now permanently installed in the two towns.

CERAMICS: A GREEN APPROACH

CONCLUSION

Hopefully this book has told you a bit about the constant development of more cost-effective ways to process raw materials and improve the footprint that the extraction of those materials creates, and encouraged you to find better ways to minimise the amount of fossil fuels used in ceramics. It is an ever-changing situation and a positive thing that the conversation has at least started about green pottery practises.

What does being green mean to potters? This will depend on the processes they use. A wood firer will defend their practise as green, while someone else would regard it as polluting. Salt, soda, and reduction firing with fossil fuels all have an impact on our environment. Be sure of your facts; don't listen to hearsay; do the research, expand your knowledge and be wary of books written with second-hand information or a personal agenda.

I personally don't trust anything until I have done it myself and seen the results. There is so much incorrect information online, presented by people who think they are experts but more often are not. It's referred to as the Dunning-Kruger effect.

Ceramics, once you have fired them, have a long lifespan – in many cases until they are broken. Even then, they can have a new life, as we have seen examples of. It comes down to how we use our finite resources sensibly. Make your work in a way that shows respect for the raw materials and the effect they have had on the environment on their long journey to your workshop or studio. Don't waste our valuable resources. Dispose of any waste in a responsible way so as not to contaminate our environment. Don't waste clean water, re-use it. We are fortunate as potters that our main ingredient, clay, is easy

to recycle at little or no cost to us or the environment.

Choose and use your fuel wisely and in the most viable and cost-effective way. Hopefully, in the not-too-distant future, we can look forward to a time when clean electricity from sustainable sources is easily available at a cost that does not make the process of firing too expensive for potters. At some point hydrogen may become easily available to fire larger kilns and we can find ways of using any waste heat, creating little or no impact on the environment.

I think it will be some time before wood disappears from the world of the studio potter, as there is an inextricable link with wood, flame and ash that is so beloved by many potters. I think it is here to stay for some time but understand that even if it might be considered carbon neutral, it is still a polluter.

Previous page: Plates by Russell Kingston. Photo © Russell Kingston.
Above: Plates by Jessica Turrell. Photo © Jessica Turrell

CONCLUSION

If salt or soda firing is your thing, make sure it is used for good reason not just because you can, and not in the hope it may save a bad pot. It never will. Question whether the quality it gives is integral to your work and that it is acceptable to both you and the environment.

The sourcing of sustainable wood and the correct preparation of the timber is something to be aware of, as is the distance it is transported. It should be dried out by natural means, not expending fuel to do so, i.e., kiln dried. If we make sure these paths are followed, it will mean this method can live on for some time.

Potters have a duty to understand the impact the materials and firing processes have on our environment, and we are honour bound to use them in the most respectful way possible and not squander them as we did in the past. Explore and research how to pack and dispatch you work in the greenest possible way and make sure you understand the impact a particular type of packing material has on the environment. Don't just assume it is green; know it is green! Do the research without compromising the safe arrival of your work and its possible return.

Know your materials and know your craft. Respect and acknowledge the earth that it came from, where it is bound to return to at some time in the future. Keep up with research on new and efficient uses of fuel, developments in kiln design and thermally efficient refectories. Think long term – not short term, just because it's a bit cheaper.

Most importantly, make great inspiring pots. It may be worth noting that historically, an apprenticeship for a thrower (not a potter) in the UK was seven years and nothing would be fired until acceptable standards were reached. Similar systems were employed in Japan. Recycle the bad pots, and only fire the good ones, so that when someone digs them up in a thousand years' time, you can still be proud of them. We never stop learning, I know I don't. You never know it all, no matter what some potters think. You never know if your pots may end up representing the art of the potter in the twenty-first century and they may end up as examples in a museum one day.

Be kind to yourself and the planet. Believe in the future of ceramics. Adopt small changes if big ones scare you. Making a difference is possible, without sacrificing your art.

CERAMICS: A GREEN APPROACH

GLOSSARY OF TERMS

Aniseed oil – medium used in on-glaze painting.

Ball clay – a fine, plastic clay, usually firing white or off-white.

Banding wheel – a turntable operated by hand and used mainly for decorating.

Bat – a plaster or wooden disk for throwing or moving pots without handling, or for drying clay.

Bat wash – a mixture of alumina and china clay used as a protective coating for kiln shelves.

Biscuit firing – the first firing of pottery before glazing.

Body – a term used to describe a particular mixture of clay, such as stoneware body or earthenware body.

Burnishing – the process of polishing leather-hard clay using smooth pebbles or spoons to achieve a shiny surface.

Calcined – the heating of a compound or oxide.

Casting – making pots by pouring liquid clay into a plaster mould.

Casting slip – a liquid slip used in the process of making objects using plaster moulds.

Ceramic – any clay form that is fired in a kiln.

China clay – kaolin, a pure white-burning, non-plastic, bodied clay used in combination with other clays or in glazes.

Clove oil – a medium used in on-glaze painting.

Claws – tool for holding ware for glazing.

Coiling – forming a pot from coils or ropes of clay.

Combing – a method of decoration using the fingers or a toothed tool to scrape a series of lines through slip.

GLOSSARY OF TERMS

Cover coat – plastic film printed over colour in transfer printing.

Crazing – the development of fine cracks caused by excessive expansion and contraction of glaze. Usually due to under firing of body.

Dipping – applying a slip or glaze by immersion.

Earthenware – pottery fired to a relatively low temperature. The body remains porous and usually requires glazing if it is to be used for domestic ware.

Elements – the metal heating coils in an electric kiln.

Enamels – low-firing commercially manufactured colours that are painted onto a fired surface and re-fired to melt them into the glaze.

Fat oil – reduced pure turpentine, a medium used in on-glaze painting.

Feathering – a decoration made by drawing a sharp tip through wet slip.

Firing – the process by which ceramic ware is heated in a kiln to clay or glaze to maturity.

Firing Cycle – the gradual raising and lowering of the temperature of a kiln.

Foot ring – a circle of clay that forms the base of a pot.

Frazzle – a fault in on-glaze transfer printing when ware is fired too fast causing the cover coat to crawl.

Glaze – a thin glassy layer on the surface of pottery.

Glazed – glaze applied but not fired.

Glaze stain – Commercially manufactured colourant added to glaze.

Glost – glaze and fired ware.

Green ware – unfired clay ware.

Grog – fired clay that is ground into particles, ranging from a fine dust to coarse sand. When added to soft clay it adds strength, resists warping and helps reduce thermal shock.

Hand building – making pottery either by coiling, pinching or slabbing.

Incise – the process of carving or cutting a design into raw a clay surface.

In glaze – decoration applied to the glazed and fired surface and then re-fired to glaze maturing temperature again.

Kaolin – see china clay.

Kidney – a kidney shapes scraper of metal, plastic, wood or rubber.

Kiln – a device in which pottery is fired.

Kiln furniture – refractory pieces used to separate and support kiln shelves and pottery during firing.

Kneading – a method of de-airing and dispersing moisture uniformly through clay to prepare it for use.

Latex – a rubber-based glue that can be used as a peelable resist when decorating pottery.

Leather hard – clay that is stiff but no longer plastic. It is hard enough to be handled without distorting but can still be joined.

Lustre – metallic salts added in a thin layer over glaze to produce a lustrous metallic finish.

Maiolica – the name given to tin glaze earthenware with decoration.

Majolica – the term given to painted, coloured glaze on earthenware.

Mould – a plaster form used with soft clay.

Onglaze colour – see enamels

Pencil industrial – potters' term for brush.

Plastic clay – clay that can be manipulated without losing its shape.

Porcelain – fine, high firing white clay that becomes translucent when fired.

Potter's plaster – used for making moulds. The plaster hardens by chemical reaction with water. Also called plaster of Paris.

Press moulding – pressing slabs of clay into or over moulds to form shapes.

Props – tubes of refractory clay used for supporting kiln shelves during firing.

Pyrometer – equipment for reading the temperature in a kiln.

Refractory – ceramic materials that are resistant to high temperatures.

Resist – a decorative medium such as wax, latex or paper used to prevent slip or glaze from sticking to the surface of pottery.

Ribs – wooden or plastic tools used to lift the walls of thrown pots.

Saggar - refractory box used to protect ware from direct flame.

Sgraffito – scratching through a layer of clay, slip or glaze to reveal the colour underneath.

Short – the term used to describe soft clay lacking in plasticity.

Slabbing – making pottery from slabs of clay.

Slip – liquid clay.

GLOSSARY OF TERMS

Slip trailing – decorating with coloured slip squeezed through a nozzle.

Soak – allowing the kiln to remain at a specific temperature for a set time.

Soft soap – a semi- liquid soap used to form a release in mould making.

Spit out or Sugaring – pitted surface on enamelled ware due to moisture escaping through the softened glaze layer.

Sponging – a decorative method of applying slip or glaze, or cleaning the surface of pottery before firing.

Spray booth – equipment for extracting dust from glaze spraying.

Sprig – a moulded clay form used as an applied decoration.

Stains – unfired colours used for decorating pottery or a ceramic pigment used to add colour to glazes and bodies.

Stalls – devise to put to put on finger ends to aid glazing.

Stilts – a small stand used to support pots in a firing to prevent glazed surfaces coming into contact with the kiln shelf.

Stoneware – vitrified clay, usually fired above 1200°C (2190° F).

Terracotta – an iron-bearing earthenware clay that matures at low temperatures and fires to a rich red colour.

Thermal shock – sudden increase or decrease in temperature that puts great stress on a fired clay body, causing it to crack.

Thermocouple – an instrument placed in a kiln to measure the temperature.

Trimming or turning – removing spare clay from the base of a thrown pot with a sharp loop tool while the pot revolves on the wheel.

Torsion viscometer – a device used to measure the fluidity and thixoprity of liquids.

Under glaze – a colour that is usually applied to either green ware or bisque fired pottery and in most cases covered with a glaze.

Vitrified – refers to clays which fire to high temperatures so that the clay particles fuse together and become glass-like.

Wax resist – the process of decorating by painting wax on a surface to resist a water-based covering.

Wedging – a method of preparing clay for use or mixing different clays together to an even, air-free consistency before kneading.

CERAMICS: A GREEN APPROACH

ADDITIONAL REFERENCES

Alumina
Increases glaze viscosity, firing range and resistance to crystallisation.

Alumina Hydrate
An alumina source rarely used in clay bodies or glazes but for kiln shelf wash, wadding, and in a granular form, as a placing sand for firing delicate items and bone china. Small additions increase the viscosity of glaze melt but should not be used as a matting agent because it produces immature glaze not suitable for functional glazing.

Barium Carbonate – Poisonous
A secondary flux in stoneware and porcelain glazes – produces vellum matt. Up to 2.5% can be added to some clay bodies to prevent scumming arising from soluble salts.

Borax – Poisonous
A vigorous low temperature glaze flux. Slightly soluble in water so usually introduced to glazes as a frit.

Calcium Chloride
Used as a flocculent in glazes having the effect of allowing the glaze constituents to settle in a loosely-packed arrangement thus making them more easily reconstituted.

Mix the flocculent in a cup with hot water until no more can be dissolved. Allow to cool then mix only a few drops per litre of glaze and mix thoroughly. Larger additions will cause the glaze to thicken – useful when applying the glaze to vitreous wares.

Colemenite (Boro–Calcite)
A useful and powerful flux used in glazes to introduce an insoluble form of boron with calcium.

Gerstley Borate
A variety of Colemenite – used as a flux in studio glazes.

Christobalite – Hazardous if inhaled
A powdered, pre-fired form of silica used to improve craze resistance of slips and bodies.

Dolomite
A naturally occurring combination of calcium and magnesium carbonates providing a secondary flux for high temperature porcelain and stoneware glazes.

Feldspar
A common and naturally occurring mineral used as the major flux in clay bodies and in high temperature glazes.

Additional References

Potash felspar (orthoclase) - The most commonly used form of feldspar.
Soda feldspar (Albite) – another form of feldspar.

Fire clay
Refractory clay used as an additive to stoneware bodies to produce an open texture and speckling (under reduction).

Flint – Hazardous if inhaled
A refractory material used to provide silica in bodies and glazes. Increases firing temperature and craze resistance but reduces plasticity and shrinkage.

Lithium Carbonate
An alkaline flux used as a substitute for potash or soda where a good craze resistance is required. It provides an alkaline colour response.

Magnesium Sulphate (Epsom salts)
Used to flocculate glazes to assist suspension and application to more vitreous wares. See Calcium Chloride.

Nepheline Syenite
A mineral mixture of feldspar and hornblende with little silica – More fusible than feldspar, it can be used as a replacement to reduce the maturing range of glazes and bodies.

Quartz – Hazardous if inhaled
A form of silica used as an alternative to flint in glazes but not an exact alternative to flint in clay bodies.

Rutile
An ore containing titanium dioxide with iron oxide used to produce a mottled buff brown colour (3-8%) especially in the presence of ilmenite. Increases the opacity of glaze and exciting effects can be achieved in combination with stains or colouring oxides.

Silica sand
Usually available in several grades Used as a grog for clay bodies or as a placing sand for firing.

Talc (French Chalk, Magnesium Silicate, Soapstone)
A secondary flux introducing magnesium and used to improve craze resistance in glazes. Also a flux for clay bodies.

Tin Oxide
The oldest and most widely used opacifier producing a soft white. Add 5-10% for opacification.

Whirler
Bench or table-top wheel for hand building and modelling.

Whiting (Chalk, Limestone, Calcium Carbonate)
The main source of calcium in glazes and extensively used as a flux in stoneware and porcelain glazes. Assists hardness and durability and in large quantities produces mattness.

Wollastonite (Calcium Silicate)
An alternative to whiting as a source of lime in stoneware glazes. Useful where pin-holing is a problem.

Zircon (Zirconium Silicate)
An ultra-fine form of zircon used as an opacifier. Add 5-8% for semi-opaque and 10-15% for fully opaque glazes. Also makes a very effective wadding.

CERAMICS: A GREEN APPROACH

WEIGHTS AND MEASURES

TEMPERATURE CONVERSION FORMULA

To convert °C into °F – Multiply by 9, divide by 5 and add 32

To convert °F into °C – Deduct 32, multiply by 5 and divide by 9

BROGNIART'S FORMULA (metric)

$$W = \frac{(L-1000) \times G}{G-1}$$

Where   W = Weight in g of dry material in 1 litre of slip
G = specific gravity of dry material
(usually 2.5 for clays)
P = litre weight in g of 1 litre of slip

BROGNIART'S FORMULA (imperial)

$$W = \frac{(P-20) \times G}{G-1}$$

Where   W = Weight in oz of dry material in 1 pint of slip
G = specific gravity of dry material
(usually 2.5 for clays)
P = pint weight in oz of 1 pint of slip

UK PINT WEIGHT

Dry content of 1 pint of clay slip at various pint weights

Slip weight (oz/pint)	Dry content (oz)	SG (specific gravity)
20	0.00	1.00
21	1.67	1.05
22	3.33	1.10
23	5.00	1.15
24	6.67	1.20
25	8.33	1.25
26	10.00	1.30
27	11.67	1.35
28	13.33	1.40
29	15.00	1.45
30	16.67	1.50
31	18.33	1.55
32	20.00	1.60
33	21.67	1.65
34	23.33	1.70
35	25.00	1.75
36	26.67	1.80
37	28.33	1.85
38	30.00	1.90
39	31.67	1.95

UK PINT TO US PINT CONVERSION FORMULA

1 UK pint = 1.2 US pints

1 US pint = 0.8 UK pints

US PINT TO LITRE CONVERSION FORMULA

1 US pint = 0.5 litres

UK PINT TO LITRE CONVERSION FORMULA

1 UK pint = 0.6 litres

TO CONVERT OUNCES AND GRAMS

1 oz = 28 g

1 g = 0.04 oz

ACKNOWLEDGEMENTS

For Kay, Laura, and Callum

This is probably the most difficult part of the book to write. How do I start to acknowledge all the people who have contributed to this book? Not just the potters mentioned but all those who have imparted their time and knowledge over the 70 years of my life? They are too many to list, and I would hate to miss anyone out. So, to the many individuals and all the studios and workshops I have had the privilege to work with, as well as the technical industry of Stoke-on-Trent, thank you.

I must also thank my parents for not giving up on me as a dyslexic, when the term was yet to be invented, encouraging me in my love of art and making things, and then having the foresight to let me go to art school at 15. My training over the next four years was from tutors who were the best in their fields. A good grounding in basic skills that have, for me, stood the test of time. Combined with the many years of life drawing that taught me to see, these are probably the most important things that contributed to my career and they gave me my first teaching job.

Last of all the students at all the art schools, colleges, polytechnics and universities I have taught in over the years. You have inspired me in ways you cannot imagine.
I just hope I have been able to return the favour.